LOURDES COLLEGE LIBRARY
3 0379 1001 4647 3

D1091332

AROMATIC PLANTS: BASIC AND APPLIED ASPECTS

WORLD CROPS:

PRODUCTION, UTILIZATION, AND DESCRIPTION

volume 7

Other volumes in this series:

1. Stanton WR, Flach M, eds: SAGO The equatorial swamp as a natural resource, 1980 ISBN 90-247-2470-8.
2. Pollmer WG, Phipps RH, eds: Improvement of quality traits of maize for grain and silage use. 1980. ISBN 90-247-2289-6.
3 Bond DA, ed: Vicia faba: Feeding value, processing and viruses. 1980. ISBN 90-247-2362-0.
4. Thompson R, ed: Vicia faba: Physiology and breeding. 1981. ISBN 90-247-2496-1.
5. Bunting ES, ed: Production and utilization of protein in oilseed crops. 1981. ISBN 90-247-2532-1.
6. Hawtin G, Webb C, eds: Faba bean improvement. 1982. ISBN 90-247-2593-3.

Aromatic Plants:

Basic and Applied Aspects

Proceedings of an International Symposium on Aromatic Plants organized by the Laboratory of Ecology, University of Thessaloniki, held in Kallithea (Chalkidiki) Greece, 14-19 September 1981

edited by

NIKOS MARGARIS
Laboratory of Ecology, University of Thessaloniki, Greece

ARTHUR KOEDAM
Department of Pharmacognosy, State University Leiden, The Netherlands

DESPINA VOKOU
Laboratory of Ecology, University of Thessaloniki, Greece

1982
MARTINUS NIJHOFF PUBLISHERS
THE HAGUE / BOSTON / LONDON

Distributors:

for the United States and Canada

Kluwer Boston, Inc.
190 Old Derby Street
Hingham, MA 02043
USA

for all other countries

Kluwer Academic Publishers Group
Distribution Center
P.O. Box 322
3300 AH Dordrecht
The Netherlands

Library of Congress Cataloging in Publication Data CIP

International Symposium on Aromatic Plants (1981 :
Kallithea, Greece)
Aromatic plants.

(World crops ; v. 7)
Includes index.
1. Aromatic plants--Congresses. I. Margaris,
Nikos. II. Koedam, Arthur III. Vokou, Despina.
IV. Aristoteleion Panepistēmion Thessalonikēs.
Laboratory of Ecology. V. Title. VI. Series:
World crops (Hague)
SB301.I57 1982 582'.063 82-12593
ISBN 90-247-2720-0

ISBN 90-247-2720-0 (this volume)
ISBN 90-247-2263-2(series)

PRINTED IN THE NETHERLANDS

CONTENTS

INTRODUCTION

The use of aromatic plants has been practised since ancient times as is evidenced by records of Chinese, Egyptian, Mesopotamian, Greek and Roman origin; recent findings in Pakistan prove that it goes as far back as 5000 years. Accordingly, the importance and value ascribed to them were always high. Either as a medicine, a foodstuff, a seasoning, a cosmetic or an element of religious rituals the aromatic plant was indispensable. It is not to be forgotten that desire for the riches of India - spices among which - and the struggle to monopolize the trade of the latter contributed to the opening of new sea routes, discovering of continents and altering the picture of the known world.

With the advent of modern civilization, characterized by scientific and technological development which diverted estimation and consumption towards artificial products, aromatic plants experienced a temporary decline of use retaining, however, their importance in sectors such as the culinary art and cosmetics industry.

This situation changed during the last decade having brought worldwide a remarkable transition towards utilization of natural products. As a consequence of this revived interest the field of aromatic plants research has expanded considerably both in scope and stature, with a number of different disciplines, such as chemistry, pharmacology, botany, ecology, etc. having directed their attention on them. In parallel, the great economic importance of aromatic crops for many countries has enhanced research on them, what is evidenced by an ever increasing stream of publications. Though the main part of these papers has been devoted to the chemical composition of their volatile oils attention has also been given to genetic and chemosystematic matters, on the possible role of volatile oils upon insects, and on animal damage. Evidently, in recent years the advances in instrumentation have opened quite new avenues and much valuable information has been gained. Thus, the scanning electron microscope has revealed a world only dreamed of several years ago while analytical tools were tremendously improved by the introduction of gas-chromatography coupled with mass-spectrometry and infrared-spectroscopy. Nevertheless, a large number of aspects of aromatic plants is unsufficiently known, e.g. there is practically no information on the relationship between them and their physical and biological environment; their pharmacological activity is still unclear; the site of synthesis of the different flavour compounds is essentially unknown.

With the goal of bringing together scientists working in different disciplines of aromatic plants research to present recent progress in their field a Symposium was organized in Halkidiki, Greece, in September 1981. The meeting was attended by approximately 130 participants from 20 countries; the present volume is a collection of papers presented in it.

Since the length of this publication did not allow inclusion of all papers the editors had to undertake the difficult task of selecting only a certain number among them. Within the compass of a single specialized volume we have tried, through the papers selected, to demonstrate the multidisciplinarity of aromatic plants research. Papers included are separated into five chapters, viz. Anatomy and Morphology, Ecology and Distribution, Chemotaxonomy, Analysis and Composition, and Production and Application. We hope that this volume, reflecting the state of the art, will prove of value to all those concerned with aromatic plants.

We would like to thank all authors who have contributed articles and also Mrs. A. Karamanli-Vlahopoulou for her patience and skill in preparing the manuscript.

Arthur Koedam

Department of Pharmacognosy
State University of Leiden
The Netherlands

Nikos Margaris
Despina Vokou

Laboratory of Ecology
University of Thessaloniki
Greece

LIST OF CONTRIBUTORS

ANAGNOSTIDIS K	Institute of Systematic Botany, University of Athens, Greece
ARGYRIADOU N	Research Laboratories, VIORYL S.A., Athens, Greece
BOSABALIDIS AM	Botanical Institute, University of Thessaloniki, Greece
BOX EO	Geography Department, University of Georgia, Athens, Georgia 30602, USA
BRIESKORN CH	Institute of Pharmacy and Food Chemistry, University of Wuerzburg, Am Hubland, D-8700 Wuerzburg, West Germany
BRUNS K	Aromatic Chemicals Laboratories, Henkel KGaA, D-4000 Duesseldorf, West Germany
CHIALVA F	Martini and Rossi SpA, Piazza L. Rossi 1, 10020 Pessione, Torino, Italy
DAFNI A	Institute of Evolution, Haifa University, Haifa, Israel
DOLHAINE H	Aromatic Chemicals Laboratories, Henkel KGaA, D-4000 Duesseldorf, West Germany
ECONOMOU-AMILLI A	Institute of Systematic Botany, University of Athens, Greece
FAHN A	Department of Botany, The Hebrew University of Jerusalem, 91904 Jerusalem, Israel
FORMACEK V	Bruker Analytische Messtechnik GmbH, Am Silberstreifen, D-7512 Rheinstetten 4, West Germany
GABRI G	Martini and Rossi SpA, Piazza L. Rossi 1, 10020 Pessione, Torino, Italy
GAŠIĆ MJ	Faculty of Sciences, Department of Chemistry, University of Belgrade, Studentski trg 16, POB 550, 11000 Belgrade, Yugoslavia
GRAVEN EH	University of Fort Hare, Alice, South Africa
GREGER H	Institut für Botanik der Universität Wien, Rennweg 14, A-1030 Wien, Austria

HEYWOOD VH — Department of Botany, University of Reading, Berks., RG6 2AS, United Kingdom

HUSAIN SZ — Department of Botany, University of Reading, Berks., RG6 2AS, United Kingdom

KAPOR S — INEP Zemun, Belgrade, Yugoslavia

KOEDAM A — Department of Pharmacognosy, State University of Leiden, The Netherlands

KOKKINI S — Institute of Systematic Botany and Phyto-geopraphy, University of Thessaloniki, Greece

KUBECZKA K-H — Department of Pharmaceutical Biology, University of Würzburg, Mittlerer Dallenbergweg 64, D-8700 Würzburg, West Germany

LIDDLE PAP — Martini and Rossi SpA, Piazza L. Rossi 1, 10020 Pessione, Torino, Italy

MARGARIS NS — Laboratory of Ecology, POB 119, University of Thessaloniki, Greece

MARKHAM KR — Chemistry Division, D.S.I.R., Private Bag, Petone, New Zealand

McARTHUR ED — USFS-INT, Shrub Sciences Laboratory, Provo, Utah 84601, USA

NARJISSE H — E.N.A., B.P. S 40, Meknes, Morocco

NOBLE P — Institute of Pharmacy and Food Chemistry, University of Wuerzburg, Am Hubland, D-8700 Wuerzburg, West Germany

PALEVITCH D — Agricultural Research Organization, The Volcani Center, Bet Dagan, Israel

PALIĆ R — Faculty of Sciences, Department of Chemistry, Kosovo University, Pristina, Yugoslavia

PAPAGEORGIOU VP — Laboratory of Organic Chemistry, College of Engineering, Aristotle University of Thessaloniki, Greece

PAPANIKOLAOU K — Institute of Systematic Botany and Phyto-geography, University of Thessaloniki, Greece

PUTIEVSKY E — Division of Medicinal and Spice Crops, Agricultural Research Organization, Newe Ya'ar, Post Haifa 31999, Israel

RAVID U — Division of Medicinal and Spice Crops, Agricultural Research Organization, Newe Ya'ar, Post Haifa 31999, Israel

STAHL E	Department of Pharmacognosy, University of Hamburg, Bundesstr. 43, D-2000 Hamburg 13, West Germany
TSEKOS I	Botanical Institute, University of Thessaloniki, Greece
ULIAN F	Martini and Rossi SpA, Piazza L. Rossi 1, 10020 Pessione, Torino, Italy
VAN DEN BROUCKE CO	Laboratory of Pharmacognosy, Van Evenstraat 4, 3000 Leuven, Belgium
VOKOU D	Laboratory of Ecology, POB 119, University of Thessaloniki, Greece
WEBER U	Aromatic Chemicals Laboratories, Henkel KGaA, D-4000 Duesseldorf, West Germany
WELCH BL	USFS-INT, Shrub Sciences Laboratory, Provo, Utah 84601, USA
WERKER E	Department of Botany, The Hebrew University of Jerusalem, 91904 Jerusalem, Israel
WHITFIELD P	University of Fort Hare, Alice, South Africa
YANIV Z	Agricultural Research Organization, The Volcani Center, Bet Dagan, Israel

CHAPTER 1

Anatomy and Morphology

ULTRASTRUCTURE OF THE ESSENTIAL OIL SECRETION IN GLANDULAR SCALES OF *ORIGANUM DICTAMNUS* L. LEAVES

A.M. BOSABALIDIS, I. TSEKOS

1. INTRODUCTION

The essential oil producing plants, especially of the Labiatae family are very common in mediterranean flora and of great interest because of their commercial value. Most aboveground plant organs are covered with many glandular trichomes secreting the essential oil. Able to secrete are only the head cells of the glandular trichomes, which are specialized cells programmed to biosynthesize and store or extrude the oily substance.

Origanum dictamnus L. (Labiatae) is an endemic plant of the island of Crete, Greece. On both leaf surfaces there are numerous glandular scales consisting of a secretory 12-celled head, a unicellular stalk and a unicellular foot also.

The purpose of this study is to give a detailed description of the secretory process in glandular scales of *Origanum dictamnus* L. and particularly to find out the intracellular sites of essential oil biosynthesis and the manner of its elimination out of the secretory tissue as well.

2. MATERIAL AND METHODS

Seeds of *Origanum dictamnus* L. were germinated in a chamber under controlled conditions (16 h light at 30°C, 8 h dark at 25°C). Small pieces of primary leaves were fixed either in a mixture of 2.5% glutaraldehyde and 2% paraformaldehyde and then postfixed in 1% OsO_4 (7), or only in 1% OsO_4 in 0.025 M phosphate buffer, pH 7.0-7.2 at 4°C. After dehydration in an alcohol series, the specimens were gradually infiltrated and embedded in Spurr's resin. Thin sections were cut on a Reichert OmU_2 microtome, stained with uranyl acetate and lead citrate (12) and examined with a Jeol 100 B electron microscope.

3. OBSERVATIONS

The differentiation of the glandular scales begins early, when the leaf is a few millimeters in length. Each scale originates from a single proto-

Margaris N, Koedam A, and Vokou D (eds.): Aromatic Plants: Basic and Applied Aspects
 ISBN 90-247-2720-0.
Printed in the Netherlands.

dermal cell , which is more voluminous than the neighbouring epidermal cells and plasma richer. The nucleus of the initial cell of the gland occupies a great part of the protoplasm and in most cases is surrounded by many organelles (Fig. 1). The stroma of the plastids is electron dense containing usually globular inclusions and few dilated thylakoids. Mitochondria are round to oval in shape and their matrix bears many sacculi and DNA-areas as well. At this developmental stage the ER is represented by some short rough cisternae scattered in cytoplasm.

As differentiation proceeds the gland initially undergoes two successive periclinal divisions giving rise to the foot cell, the stalk cell, and the head mother cell. During the first anticlinal divisions of the head mother cell, the number of mitochondria increases and progressively the inclusions disappear from the plastid stroma (Fig. 2). The plastids of the stalk and foot cell retain their inclusions until senescence. The separating walls of the head cells are thin and contain numerous plasmodesmata.

After the divisions have been completed and the protoplasm of the gland cells is well organized, the head cells of the scales enter the secretion phase. A striking feature at the onset of secretion is an observed asymmetry in the two dark layers of the plasmalemma lining the apical walls (Fig. 3, arrowheads). Although at this developmental stage there is no evident presence of osmiophilic material in the cytoplasm of the secretory cells, the upper walls of the head appear impregnated with an electron dense substance. This substance is eliminated into a space, which is formed by detachment of the cuticle from the walls (Fig. 3). During the advance of the secretory process the volume of this subcuticular space progressively increases (continuous discharge of the exudate into it) and many mitochondria accumulate along the plasmalemma (Fig. 4). It must be mentioned in addition, that only the apical periclinal walls of the head cells and none of the separating anticlinal walls appear impregnated with the secretory product (Fig. 4, single and double arrow).

At the stage of the active secretion many small osmiophilic droplets are observed in the cytoplasm among various organelles (Fig. 5). They do not possess any detectable boundary membrane and gradually fuse with each other forming larger droplets (Fig. 6). During this process the cytoplasm loses progressively its initial density and a great part of its mass appears electron transparent (Figs. 5, 6) . The plastids do not change significantly in number and their stroma has a fine granular structure with no inclusions at all. Mitochondria are numerous, round in shape and dispersed throughout the cytoplasm.

Beside the osmiophilic droplets one also encounters many electron translucent cytoplasmic droplets, some of which appear to discharge into vacuoles (Figs. 7, 9).

The population of both kinds of lipophilic droplets occupies a considerable part of the secretory cells interior, and moves exclusively towards the apical walls of the head. It is at these positions they come into contact with the plasmalemma, which exhibits the above-mentioned structural asymmetry.

The passage through the plasmalemma and the wall microfibrils is usually connected with an alteration in the structure and electron density of the secretory product. In this sense the homogenous osmiophilic droplets become less electron dense and fine-granular after they have passed the plasmalemma (Fig. 8), while the electron translucent ones change to a rough-granular and electron opaque substance (Fig. 9).

Presence of lipophilic material with morphology corresponding to that of the cytoplasmic droplets was not observed within the various organelles and particularly the plastids and the ER-elements. Only some vacuoles appear to contain such droplets in the vicinity of the tonoplast (Figs. 9, 10).

In a more advanced developmental stage the head cells of the glandular scale exhibit an increased electron density and their anticlinal walls become refolded under the sufficient pressure exerted on them by the subcuticular space, which is full of oil. At this stage the secretory cells begin to present the first signs of disintegration referring to the various organelles and other cytoplasmic membranes. The nucleus has a flake-like structure and the nuclear envelope has been destroyed (Fig. 11). The nucleolus has entirely lost its granular appearance and it is now homogenous with a median electron density. The matrix-sacculi system of mitochondria retains its original structure no more turning partly into a fluid, which is discharged into the cytoplasm (Fig. 14). The mitochondrial envelope becomes locally dilated or its continuity is interrupted by small accumulations of a substance probably deriving from the disintegration of the biomembranes (Fig. 13, arrowheads). Analogous features are also observed in tonoplast (Fig. 12, arrowheads), and may be considered as a general characteristic of the endomembranous system at this stage.

By the end of the life cycle of the secretory cells, the intracellular space possesses no detectable structure and a great part of it is occupied by a homogenous substance identical in appearance to the content of the subcuticular space (Fig. 15).

4. DISCUSSION

The most striking feature in glandular scales of *Origanum dictamnus* L. at the onset of secretion is an observed asymmetry in plasmalemma lining the apical walls of the head. Asymmetrical structure of plasmalemma has been suggested by Schnepf (13, 14, 15) to be connected to an active passage of the exudate through this membrane mainly in form of molecules (eccrine secretion).

The apical walls of the secretory cells appear at this stage impregnated with the osmiophilic oil, while at the same time no detectable presence of a morphologically identical substance was observed in the cytoplasm. This event probably suggests, that the exudate is represented, at least at this stage, by its molecular form bearing non-osmiophilic characteristics. The osmiophilia, is obtained after the exudate has passed the plasmalemma (18).

In the progress of secretion, droplets of essential oil having different size and electron density are formed in the head cells interior. Varying in electron density of droplets is probably due to a different constitution of droplets (final mixture of individual oil components takes place within the subcuticular space). These droplets are believed to be synthesized in cytoplasm (without excluding the possibility of an indirect implication of various organelles to this function). This speculation is based on the following observations:

a. As the number of cytoplasmic droplets increases, the density of the cytoplasm becomes reduced and many electron transparent regions are formed among positions. Such an image strongly suggests that the production of the secretory droplets takes place at the expense of the cytoplasm.

b. Lipophilic droplets have never been located at the stage of secretion inside any organelle and particularly the plastids and ER-elements, which constitute the main sites of terpenoid biogenesis (16,17, 18, 20). Also a releasing of a fluid identical to the cytoplasmic droplets from the interior of the above-mentioned organelles into the cytoplasm was not observed.

c. Cytoplasmic droplets are not limited by a membrane, showing they are not derived from any organelle.

d. Droplets are dispersed in cytoplasm, round in shape and of various size.

A cytoplasmic origin of terpenoids has been reported by many authors both at light (2, 9, 10) and electron microscope level (4, 6, 11). Dell and McComb (3) in studying the glandular trichomes of *Newcastelia* using radio-active material, concluded that the site of terpene biosynthesis is located in the cytoplasm of the secretory cells.

Small cytoplasmic droplets become larger by fusing without the production of new ones to be ceased. The so-formed population of essential oil droplets exhibits a varied osmiophilia and progressively moves towards the top walls of the head, where it contacts the plasmalemma. Any lack of a detectable membrane limiting the droplets suggests that the releasing of the secretory substance into the wall is performed not by way of membrane coalescence, but exclusively throughout the plasmalemma (the indispensable energy to this function is provided by the mitochondria accumulated along the plasmalemma). This event presupposes a special structure of the plasmalemma and a suitable enzymic equipment so that large amounts of essential oil could pass through this plasma membrane, without causing any damage to it.

The intraplasmic movement of the secretory droplets takes place exclusively towards the head apex and not to any lateral direction. This results from the fact that only the upper walls of the head and none of the anticlinal separating walls are impregnated with the exudate. The latter apart from being free of exudate impregnation, also lack plasmodesmata, showing that head cells secrete independently from each other (no apoplastic or symplastic transportation of secretory material). In addition only the plasmalemma lining the apical walls of the head demonstrates an asymmetrical structure.

This one-direction movement of cytoplasmic droplets is presumably associated with the cell polarity within the scope of which the presence, in top wall plasmalemma, of a "pump mechanism" drawing the droplets towards the head apex could be assumed.

Such an eccrine secretory process is also observed in case of other lipophilic substances as suberin (1), sporopollenin (19), cholesterol (5), etc.

After the exudate has passed the plasmalemma and the wall, it is discharged into a space, which is formed by detachment of the cuticle from the outer wall surface. The opening of this space is due to the fact, that the cuticle is free of pores and on the outside covered by a thin layer of wax.

At the postsecretory period the organelles of the secretory cells become deformed and later destroyed, while at the same time a great part of the protoplasm is occupied by a substance identical in appearance to the subcuticular space content. This observation is in accordance with the view of Loomis and Croteau (8), who consider that mass production of essential oil coincides with the phase of disintegration of the secretory cells endomembranous system.

REFERENCES

1. Barkhausen R, Rosenstock G. 1973. *Z. Pflanzenphysiol.* 69:193.
2. Behrens J. 1886. *Ber. Dtsch. Bot. Ges.* 4:400.
3. Dell B, McComb AJ. 1978. *J. Exp. Bot.* 29:89.
4. Dickenson PB, Fairbairn JW. 1975. *Ann. Bot.* 39:707.
5. Elliot CG, Knights BA. 1974. *Bioch. Biophys. Acta.* 360:78.
6. Heinrich G. 1970. *Protoplasma* 70:317.
7. Karnovsky MJ. 1965. *J. Cell Biol.* 27:137A.
8. Loomis WV, Croteau R. 1973. In: Recent Advances in Phytochemistry. Runeckles JC, Mabry TJ (eds.), New York, Academic Press.
9. Middendorf E. 1927. *Beitr. Biol. Pflanzen* 15:61.
10. Müller R. 1905. *Ber. Dtsch. Bot. Ges.* 23:292.
11. Rachmilevitz T, Joel DM. 1976. *Isr. J. Bot.* 25:127.
12. Reynolds ES. 1963. *J. Cell Biol.* 17:208.
13. Schnepf E. 1964. *Protoplasma* 58:193.
14. Schnepf E. 1965. *Z. Pflanzenphysiol.* 53:245.
15. Schnepf E. 1969. Sekretion und Exkretion bei Pflanzen. Protoplasmatologia, Bd. 8, Wien, New York, Springer-Verlag.
16. Schnepf E. 1969. *Protoplasma* 67:185.
17. Schnepf E. 1969a. *Protoplasma* 67:195.
18. Schnepf E. 1969b. *Protoplasma* 67:205.
19. Vasil IK, Aldrich HC. 1970. *Protoplasma* 71:1.
20. Wooding FBP, Northcote DH. 1965. *J. Ultr. Res.* 13:233.

EXPLANATION OF FIGURES

Abbreviations:Cu=cuticle, CW=cell wall, Cy=cytoplasm, ER=endoplasmic reticulum, FC=foot cell, HC=head cell, M=mitochondrion, N=nucleus, Nu=nucleolus, OD=oil droplet, P=plastid, Pd=plasmodesma, PL=plasmalemma, SC=stalk cells, SCS=subcuticular space, T=tonoplast, V=vacuole.

FIGURE 1. The initial cell of a glandular scale. Many organelles are arranged around the nucleus (N), which occupies a great part of the cell interior. Note the density of the cytoplasm and the presence of globular inclusions in plastid stroma (P).
FIGURE 2. A 6-celled differentiation stage of a glandular scale sectioned longitudinally. Plastids vary in form and size and together with mitochondria are scattered throughout the cytoplasm. The anticlinal walls of the head cells are very thin and contain plasmodesmata (Pd).
FIGURE 3. Portion of the plasmalemma (Pl) lining the apical walls of the head. Arrowheads indicate asymmetrical structure of the unit membrane. Cell wall (CW) appears impregnated with an osmiophilic substance, which is accumulated in a space (SCS) formed by detachment of the cuticle (Cu) from the wall.
FIGURE 4. Two adjacent secretory cells of the head. Apical walls appear impregnated with the osmiophilic exudate, while no presence of an identical substance is located in the anticlinal separating walls (single and double arrow). Many mitochondria (M) are aggregated in upper cytoplasm. Black dash-line separates cell wall (CW) from subcuticular space (SCS).
FIGURE 5. Stage of active secretion. At this stage many osmiophilic droplets (OD) of various size are formed in the cytoplasm of the secretory cells. Compare cytoplasm density with that of the first differentiation stages (Figs. 1, 2).
FIGURE 6. Large secretory droplets (OD) in cytoplasm among organelles. As droplets become larger the density of the cytoplasm is strongly reduced and many electron transparent regions are formed.
FIGURE 7. Portion of protoplasm in a secreting head cell. Besides osmiophilic droplets one also encounters many electron translucent droplets (OD).
FIGURES 8 and 9. Both kinds of secretory droplets (OD) migrate towards the apical walls of the head and contact the plasmalemma (PL). Note exudate impregnation of the cell wall and presence of a fine-granular substance within the subcuticular space (SCS).
FIGURE 10. Intravacuolar osmiophilic globule (OD) morphologically identical to the content of the subcuticular space (SCS).
FIGURES 11-15. Degenerating secretory cells.
FIGURE 11. Destroyed nuclear envelope is illustrated by black dash-line. Note the appearance of the karyolymph and the nucleolus (Nu). Cell wall refoldings also can be seen (CW).
FIGURE 12. Portion of a vacuole limited by degraded tonoplast (T). The continuity of the tonoplast appears interrupted by small accumulations of an electron translucent substance (arrowheads).
FIGURE 13. Aggregation of mitochondria (M) showing degradation in boundary envelope (arrowheads).
FIGURE 14. Region of cortical cytoplasm containing a disintegrated mitochondrion. An electron translucent fluid appears to be released from the mitochondrial matrix into the cytoplasm.
FIGURE 15. Secretory cell at collapse. Extended areas of cell interior are occupied by an osmiophilic substance resembling the subcuticular space content.

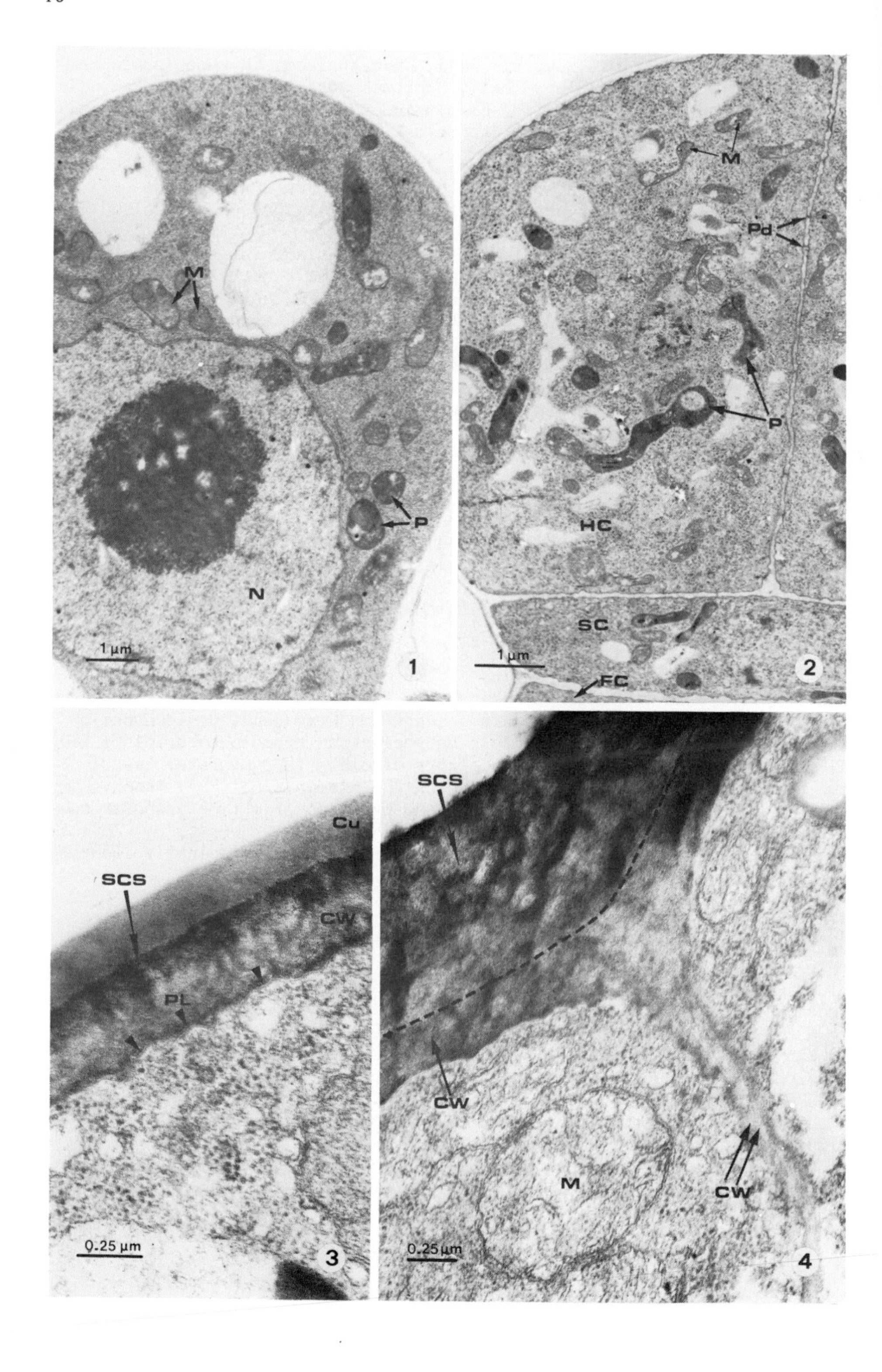
M
P
N
1 µm
1
M
Pd
P
HC
SC
FC
1 µm
2
Cu
SCS
CW
PL
0.25 µm
3
SCS
CW
M
CW
0.25 µm
4

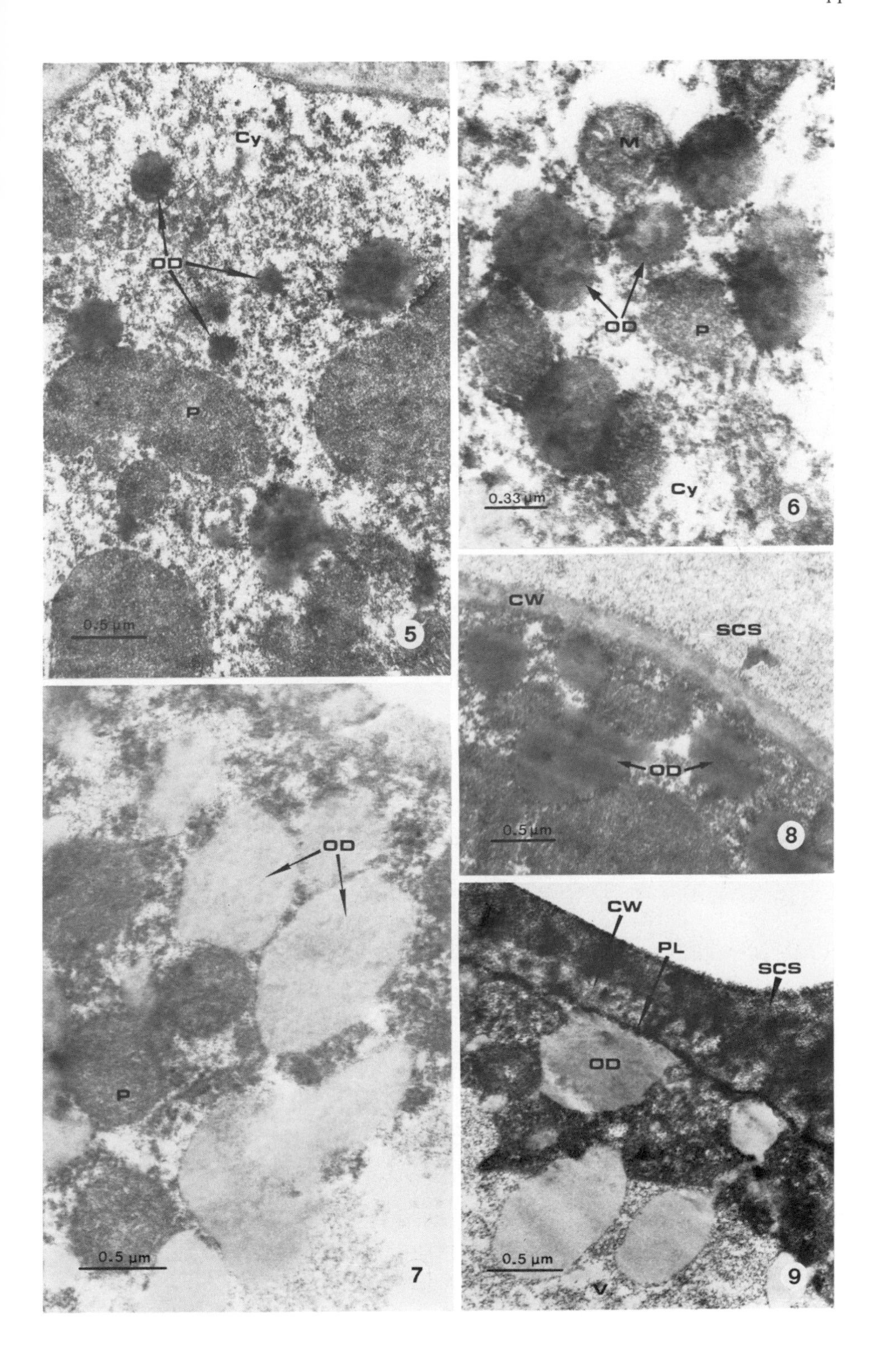
Cy
OD
P
0.5 µm
5
M
OD
P
Cy
0.33 µm
6
OD
P
0.5 µm
7
CW
SCS
OD
0.5 µm
8
CW
PL
SCS
OD
V
0.5 µm
9

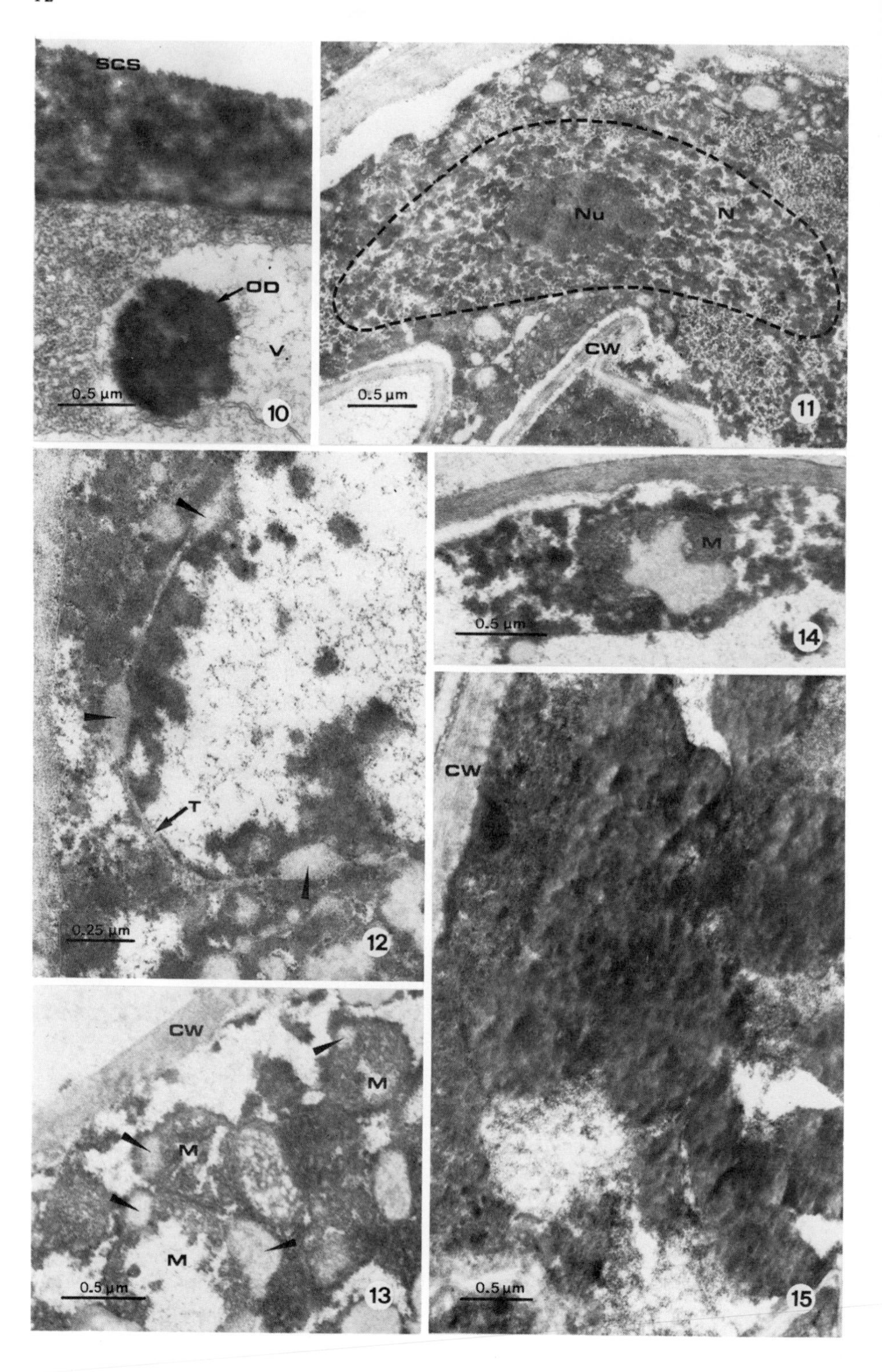
SCS
OD
V
0.5 µm
10
Nu
N
CW
0.5 µm
11
T
0.25 µm
12
M
0.5 µm
14
CW
M
M
M
0.5 µm
13
CW
0.5 µm
15

LEAF MORPHOLOGY OF *THYMUS CAPITATUS* (LABIATAE) BY SCANNING ELECTRON MICROSCOPY

A. ECONOMOU-AMILLI, D. VOKOU, K. ANAGNOSTIDIS, N.S. MARGARIS

1. INTRODUCTION

Thymus capitatus (L.) Hoffmann & Link, a common mediterranean plant of Labiatae (12), rich in essential oil (5), represents in Greece an important element of the phryganic flora.

The chemical composition of its essential oil has been thoroughly investigated (7, 8, 9, 10).

In morphology, this aromatic plant is a dwarf shrub with ascending to erect woody branches bearing axillary leaf-clusters. Its seasonally dimorphic leaves, small in the dry and large in the wet season (DSL and WSL, respectively) are of long shoots, 6-10 X 1-1.2 mm, sessile, linear acute, almost trilateral, sparsely ciliate at the base; the corolla is typical of the family of Labiatae, with the upper lip bifid (12).

In this paper, the external leaf morphology of *Thymus capitatus* is examined with the scanning electron microscope. Emphasis is given on some differentiated epidermic structures of the leaf, such as glands, stomata and trichomes (hairs). Apart from morphology, attention has been also given to the distribution and frequency of these structures. For this purpose, there were examined both the upper and the lower surfaces of DSL and WSL and of irrigated and non-irrigated plants. It is worth noticing that representatives of Labiatae have partly been investigated with SEM (2, 3, 6) while concerning *Thymus capitatus* related data are not available.

2. MATERIAL AND METHODS

Both irrigated and non-irrigated plants of *Thymus capitatus* were collected from Mt. Hymettus, Kessariani (Athens) in the vicinity of the University Campus, in June 1981, when both types of leaves coexist on the plant. For scanning electron microscopy, specimens were examined from material dried in air or previously rinsed with distilled water. The damage of the leaves resulting from this procedure is rather insignificant, since the nature of the material - hard and dry surfaces - permits use of techniques not re-

Margaris N, Koedam A, and Vokou D (eds.): Aromatic Plants: Basic and Applied Aspects

ISBN 90-247-2720-0. Printed in the Netherlands.

quiring gradual dehydration. All electron micrographs were taken on a JEOL-JSM-35 using Agfa 25 Professional film from specimens coated with 1500 Å gold palladium using JFC-1100 Ion Sputter.

3. REMARKS

In leaf clusters of *Thymus capitatus*, the trilateral nature of leaves is obviously seen under SEM even in low magnifications (Fig. 2.1). The dorsal part (lower surface of leaves) is characterized by a middle ridge (Figs. 2.2, 2.3, 2.9, 2.11) decorated with short unicellular trichomes (Fig. 2.6). At both inclined sites of the dorsal part stomata and glands are rather evenly distributed (Fig. 2.3). In the ventral part of the leaves, the appearance of the epidermic structures is more or less the same but without any median interruption (Fig. 2.7).

3.1. Glandular trichomes

The epidermal glands of *Thymus capitatus*, belong to the glandular trichomes, consisted of a basic cell, a uniseriate stalk of one cell long and a head from several secretory cells (4). The cell wall around the secretory cells is differentiated into a cuticle, cuticular layer, a pectic layer and a cellulotic layer. At the stage of secretion, the cytoplasm retracts irregularly from the cell wall which becomes split between the pectic and cuticular layers, resulting in the characteristic subcuticular space where the secreted substance is accumulated (Fig. 1).

The glandular trichomes of the specimens observed appear generally isolated and only scarcely in groups of three (Fig. 2.12). They are more or less evenly distributed at both surfaces of the leaf and at both sites of the dorsal ridge. This arrangement is much more discrete in the large wet

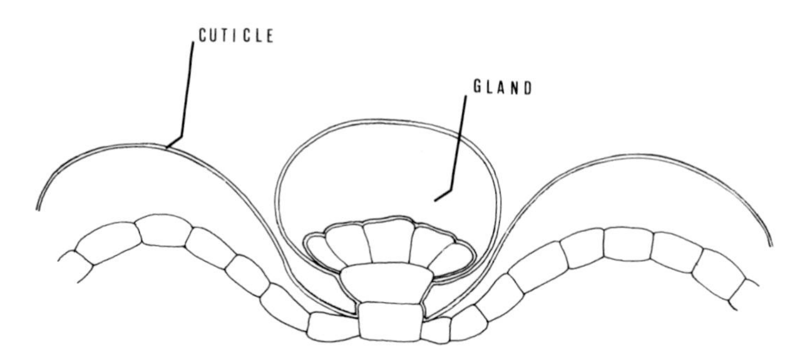

FIGURE 1. Diagrammatic representation of a glandular trichome of *Thymus capitatus* (after Fahn, 1974).

season leaves (Fig. 2.3). All glands seem to be sessile obviously due to the short stalk. The cuticle which covers the secretory cells appears either stretched (swollen), or wrinkled - in most cases - probably due to the SEM technique applied without previous gradual dehydration. It is also possible that some capitate swellings on the glands surface (Fig. 2.10, 2.11) might have the same cause, unless they are connected with functional procedures since pores in the cuticle open and droplets start to appear on the surface when pressure reaches a certain value in the subcuticular space. In some cases the direction of wrinkles observed might indicate the arrangement of the underneath secretory cells (Fig. 2.4).

Cuticle rupture was not observed and secretory cells without the surrounding cuticle have not been seen by means of a special treatment. This might indicate that the essential oil in *Thymus capitatus* remains in the subcuticular space and undergoes there a procedure of physicochemical changes, as in most Labiatae (11). It is noted here that the complete or partial evacuation of the subcuticular space as well as the way according to which it happens are always in sequence with the nature of the concentrated substance of secretion and the ecological part which the last will play (1).

To estimate the frequency of glands and make comparisons, measurements were made on: the upper and lower surfaces, of DSL and WSL, and of irrigated and non-irrigated plants. There is some evidence that glands are more numerous on the lower surface of both types of leaves and that their number in the same unit area is higher in the winter leaves. As far as irrigated plants are concerned, comparison of data did not succeed in revealing relations. In any case, results at present are not sufficient to draw a firm conclusion and need further establishment.

Additional epidermal glandular trichomes, projecting over the leaf surface with a visible stalk and involved probably in secretion, have also been observed (Figs. 2.13, 2.14). Generally, they differ in morphology as well as in the way of evacuation usually appearing with ruptured head.

3.2. Stomata

The continuity of the epidermis is interrupted by stomata, with kidney-shaped guard cells (Fig. 2.18). It is postulated that the size of the aperture between the guard cells increases or decreases as guard cells undergo changes in form, and in surface view the stoma when open appears rounder than when closed (Figs. 2.17, 2.20). The type and degree of stomatal aperture in our material might be affected by the technique applied for the preparation and observation of specimens. In *Thymus capitatus*, the type of stomata based on the arrangement of the epidermal cells neighbouring the guard cells, is diacytic (Margaris, unpublished data).

Concerning distribution and frequency of stomata in our material, it is worth noticing, although expected that there is a great difference comparing DSL and WSL. On the epidermis of the large WSL stomata are countable, markedly seen and distributed with no particular order (Fig. 2.7), while on the small DSL stomata are diminished becoming almost uncountable, or sometimes hardly visible due to the developed trichomes and wrinkles (Fig. 2.8).

3.3. Trichomes

Three types of trichomes can be distinguished on the leaves of *Thymus capitatus*. The unicellular trichomes, covered with small granules (Fig. 2.21) and varying in length, distributed at the greatest part of the leaf surface; in some regions of the ventral part they obscured stomata and made them invisible. A bundle of two-celled trichomes was markedly seen at the base of the ventral part of WSL (Figs. 2.22 - 2.24), while isolated two-celled trichomes were scattered mainly on leaf margins (Fig. 2.25) mixed with long multi-celled trichomes (Figs. 2.26-2.28). Attention was given to trichomes since they are important elements in the classification of genera and species and in analyzing interspecific hybrids (4).

ACKNOWLEDGEMENT

The second author wishes to express her thankfulness to A. ONASIS FOUNDATION for the financial aid the institution has offered to her.

REFERENCES

1. Bosabalidis AM. 1981. Light and electron microscope study on the ontogeny of *Origanum dictamnus* L. glandular trichomes and of *Citrus deliciosa* Ten. oil cavities (in Greek). Ph.D. Thesis, University of Thessaloniki, Greece.
2. Botha PMM, Black MW. 1978.(17th Annual Conference of the Electron Microscopy Society of Southern Africa, Pretoria, South Africa, Dec. 1978) *Elektronmikroskopiever suidelike Afr. Nerrigt* 8:97.
3. Croteau R, Felton M, Karp F, Kjonaas R. 1981. *Plant Physiol.* 67:821.
4. Fahn A. 1974. Plant Anatomy. 2nd ed. Oxford, New York, Toronto, Sydney, Braunschweig, Pergamon Press.
5. Hegnauer R. 1966. Chemotaxonomie der Pflanzen, Bd. 4. Basel, Stuttgart,

Birkhäuser Verlag.
6. Lovett JV, Speak MD. 1979. *Weed Res.* 19:359.
7. Papageorgiou VP. 1980. *Planta Med. Suppl*:29.
8. Papageorgiou VP, Argyriadou N. 1981. *Phytochemistry* 20:2295.
9. Philianos SM, Andriopoulou-Athanasoula T. 1977. *Biol. Gallo-Hell.* 7:93.
10. Sendra JM, Cuñat P. 1980. *Phytochemistry* 19:89.
11. Tunmann O. 1913. Mitteilungen aus der Pflanzenmikrochemie. Pharm. Post. (cited by Bosabalidis, 1981).
12. Tutin TG, Heywood VH, Burges NA, Moore DM, Valentine DH, Walters SM, Webb DA (eds.). 1972. Flora Europea. Vol. 3. Diapensiaceae to Myoporaceae. London, New York, Cambridge Univ. Press.

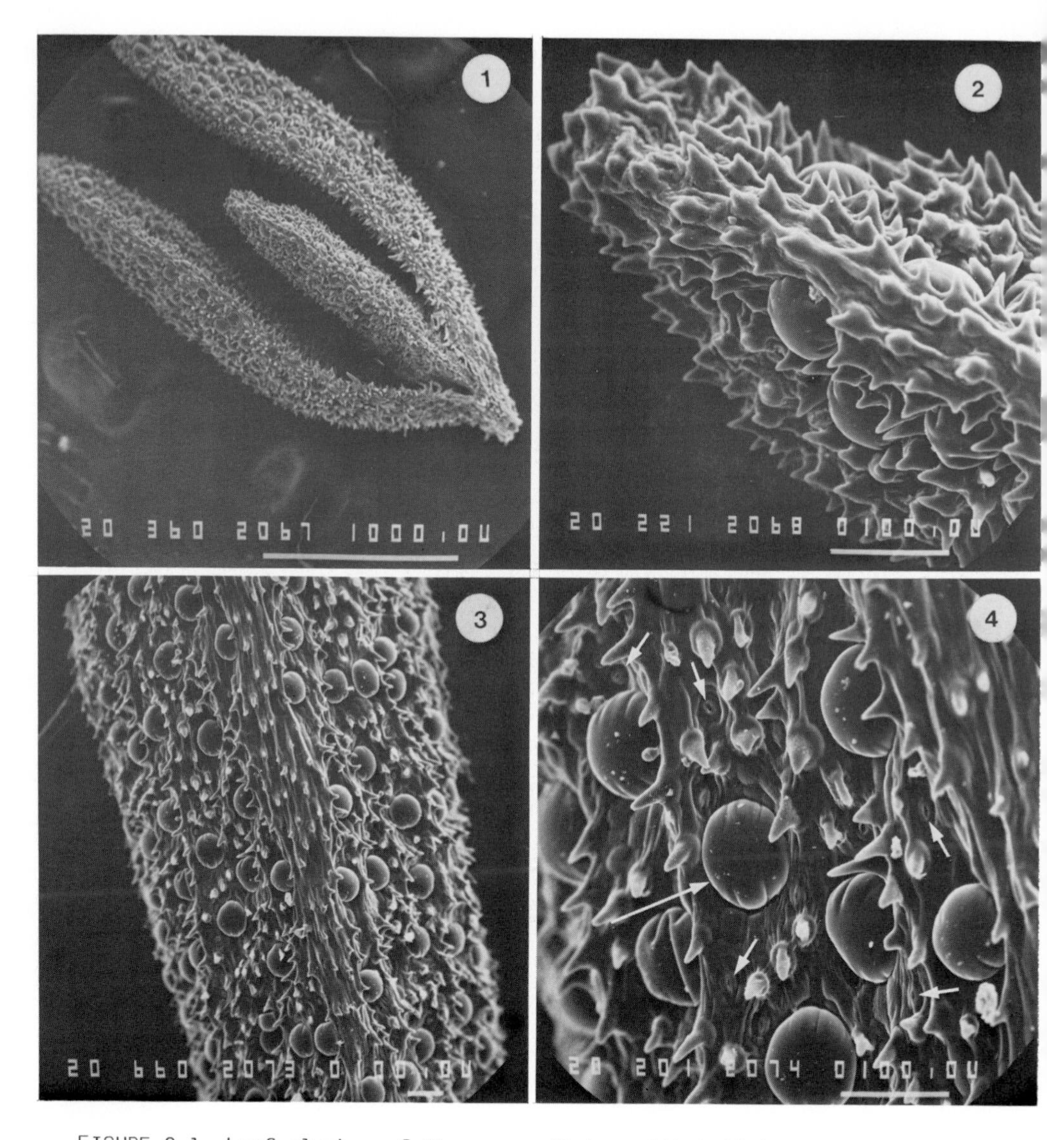

FIGURE 2.1. Leaf cluster of *Thymus capitatus* with trilateral DSL; dorsal part.

FIGURE 2.2. Upper part of a median WSL; its trilateral nature is markedly seen.

FIGURE 2.3. Median part of the same leaf; note dorsal ridge ornamented with short unicellular trichomes, and glands evenly distributed at both inclined sites.

FIGURE 2.4. Part of the inclined site of a WSL with stomata (small arrows) and wrinkled glands (large arrow).

The printed number (1000, 100, 10) in the right bottom-corner of each figure represents the scale in µm.

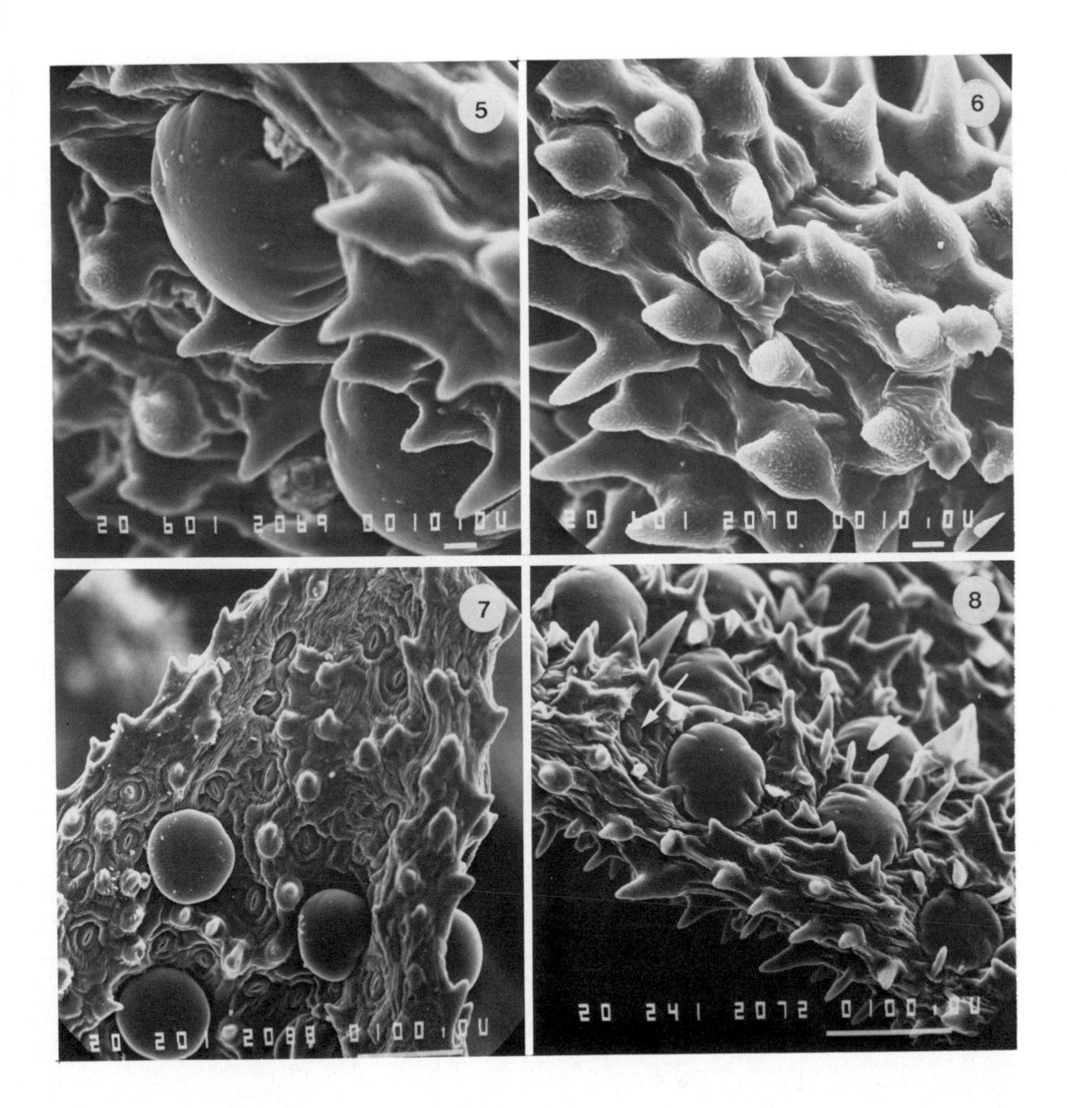

FIGURE 2.5. Glands surrounded by unicellular trichomes.

FIGURE 2.6. Unicellular trichomes on the dorsal ridge.

FIGURE 2.7. Ventral part (lower surface) of a WSL; note increased frequency of stomata compared with the same surface of a DSL (next figure).

FIGURE 2.8. Ventral part of a DSL; a stoma at arrow.

The printed number (1000, 100, 10) in the right bottom-corner of each figure represents the scale in μm.

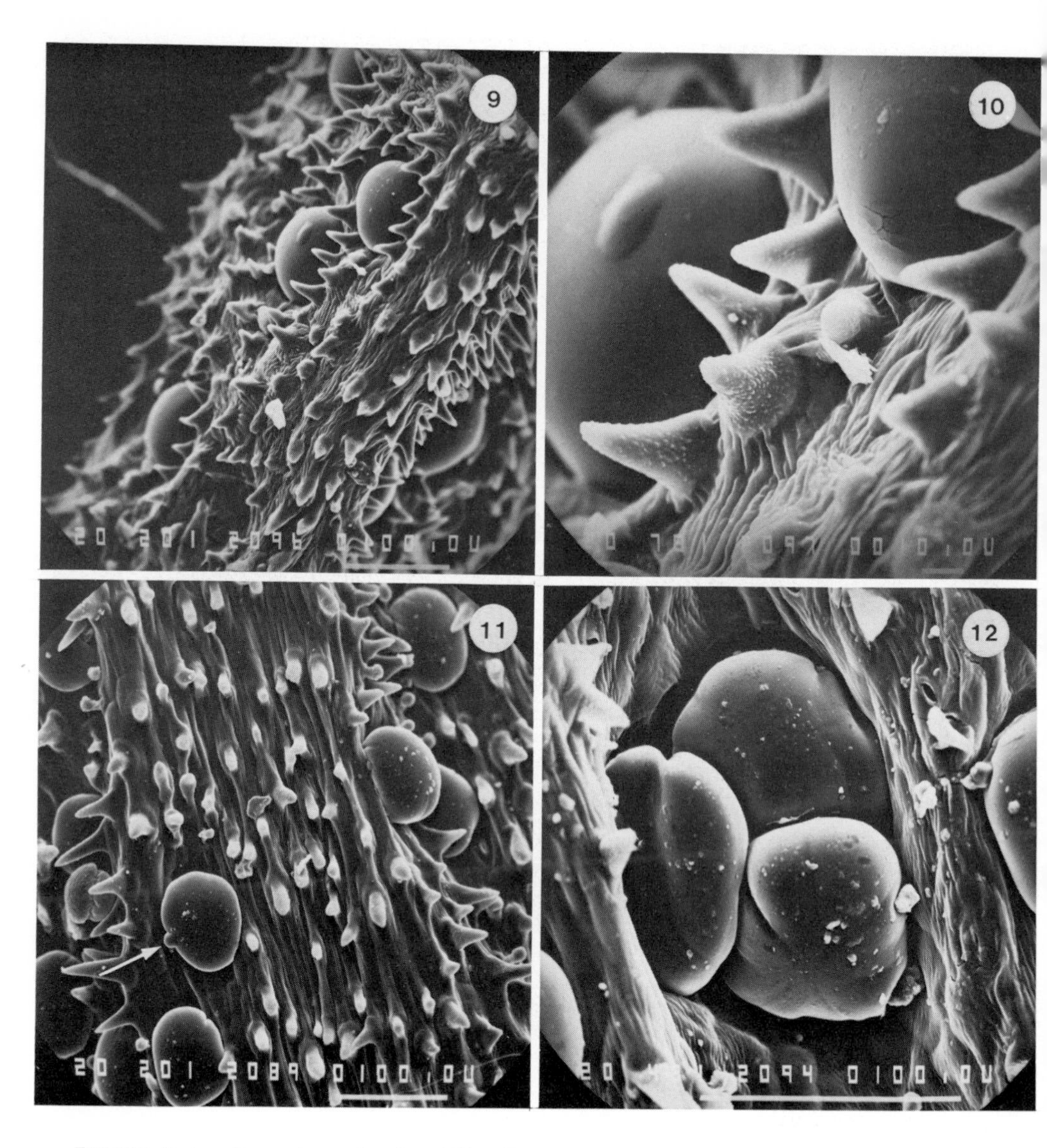

FIGURE 2.9. Dorsal part of a WSL. The glands appear swollen.

FIGURE 2.10. Gland with a swelling on the surface.

FIGURE 2.11. Dorsal part of a DSL; the arrow indicates a capitate swelling on a gland.

FIGURE 2.12. Glands appearing in group of trees (uncleaned material).

The printed number (1000, 100, 10) in the right bottom-corner of each figure represents the scale in µm.

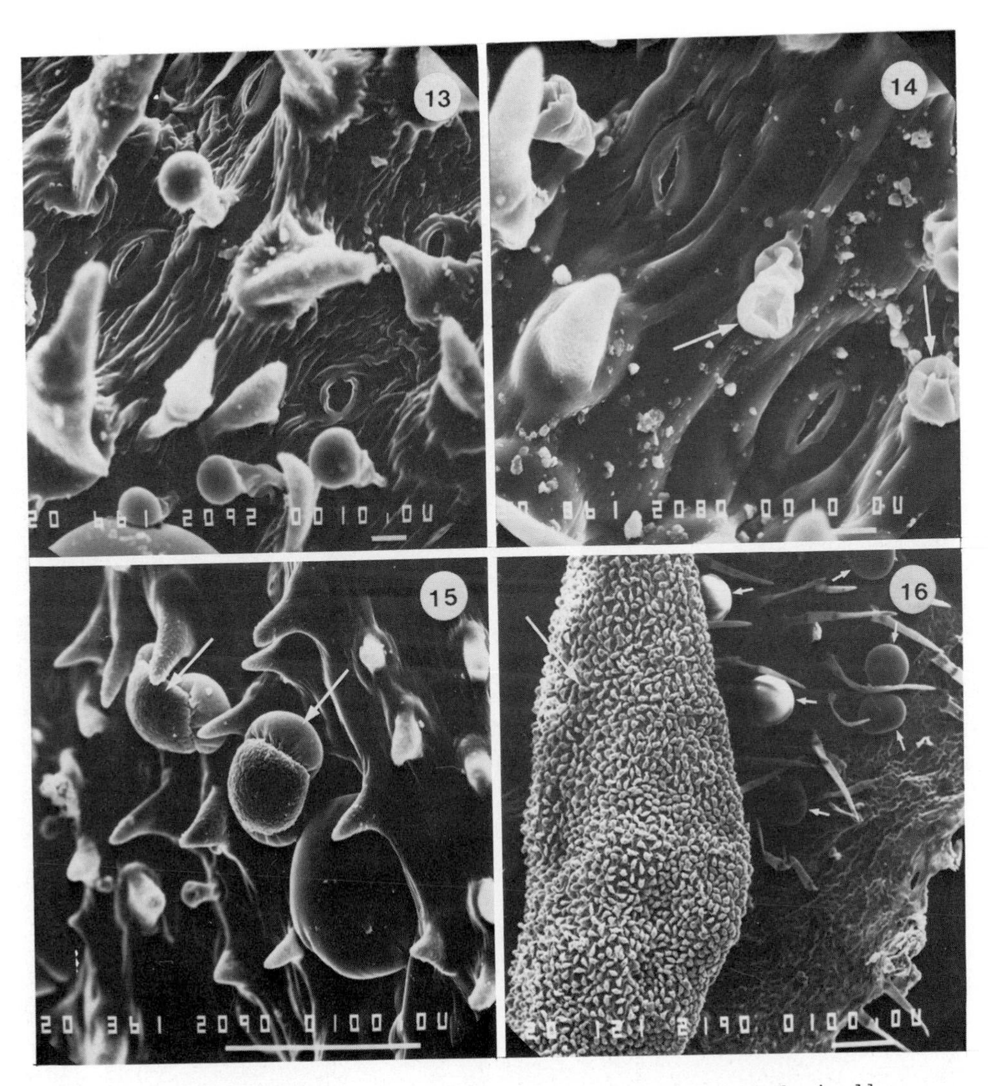

FIGURE 2.13. Epidermal glandular trichomes different morphologically (arrows); note visible stalk and smaller size.

FIGURE 2.14. The same glandular trichomes with ruptured head (arrows).

FIGURE 2.15. Unidentified structures on the leaf surface (arrows).

FIGURE 2.16. Glands on the peripheral part of the lower surface of the corolla (small arrows); refolding of the marginal part of the corolla at large arrow.

The printed number (1000, 100, 10) in the right bottom-corner of each figure represents the scale in μm.

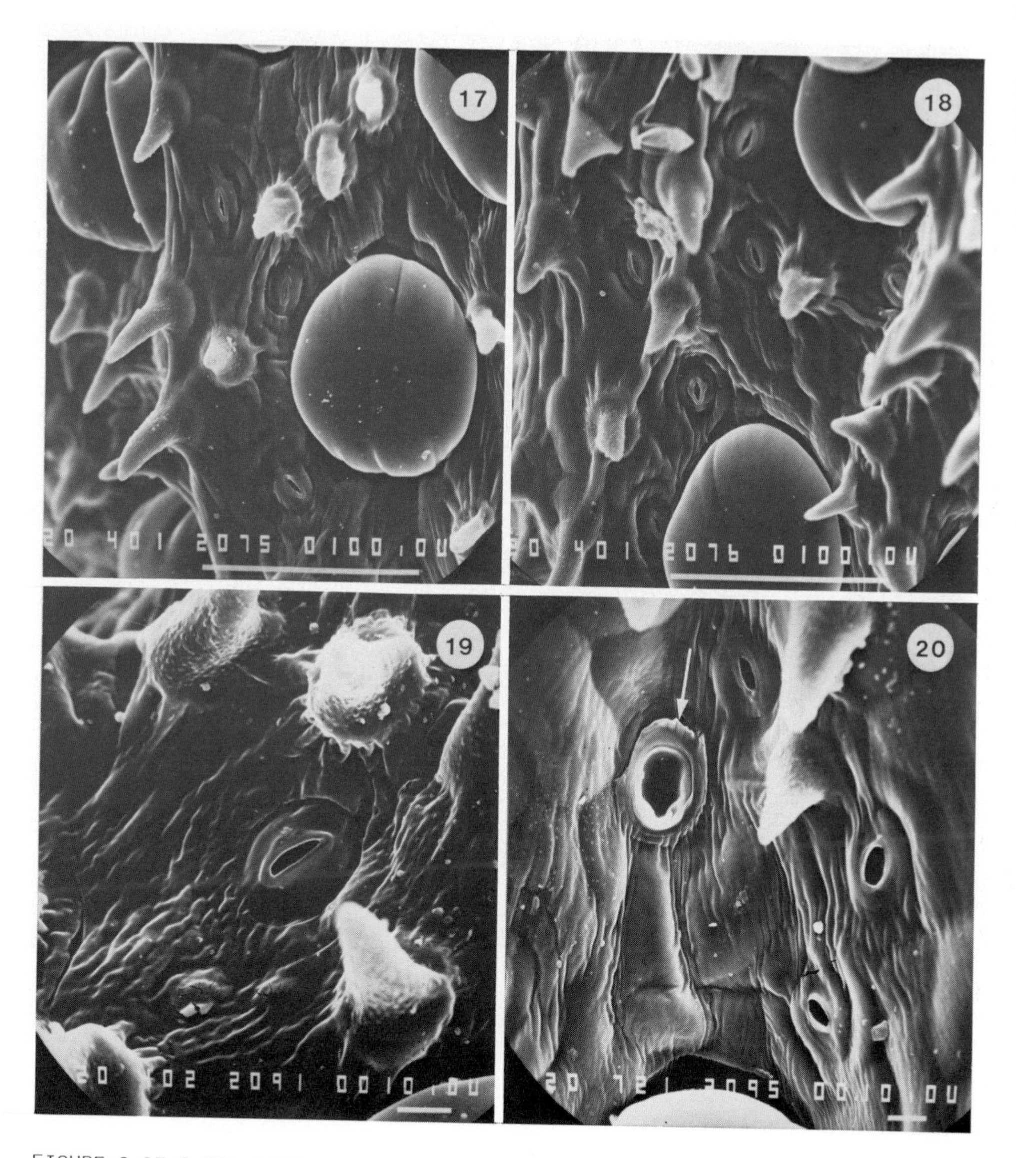

FIGURE 2.17-2.20. Different size of stomatal aperture. The arrow indicates an abnormal stoma, probably an artifact.

The printed number (1000, 100, 10) in the right bottom-corner of each figure represents the scale in μm.

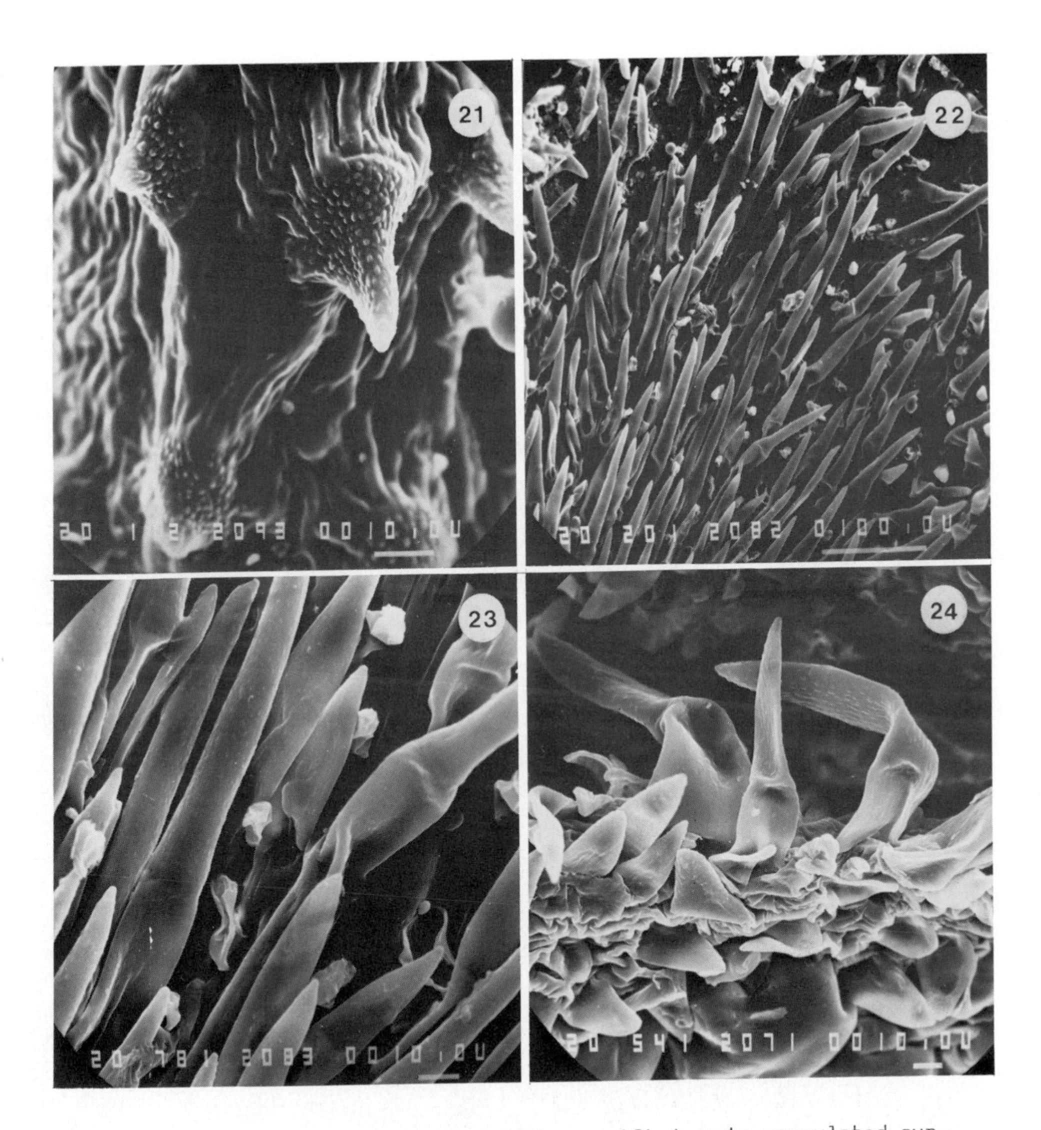

FIGURE 2.21. Unicellular trichomes highly magnified; note granulated surface.

FIGURE 2.22. Base of the ventral part of a WSL with a bundle of two-celled trichomes.

FIGURE 2.23-2.24. The same as Fig. 22; higher magnification.

The printed number (1000, 100, 10) in the right bottom-corner of each figure represents the scale in µm.

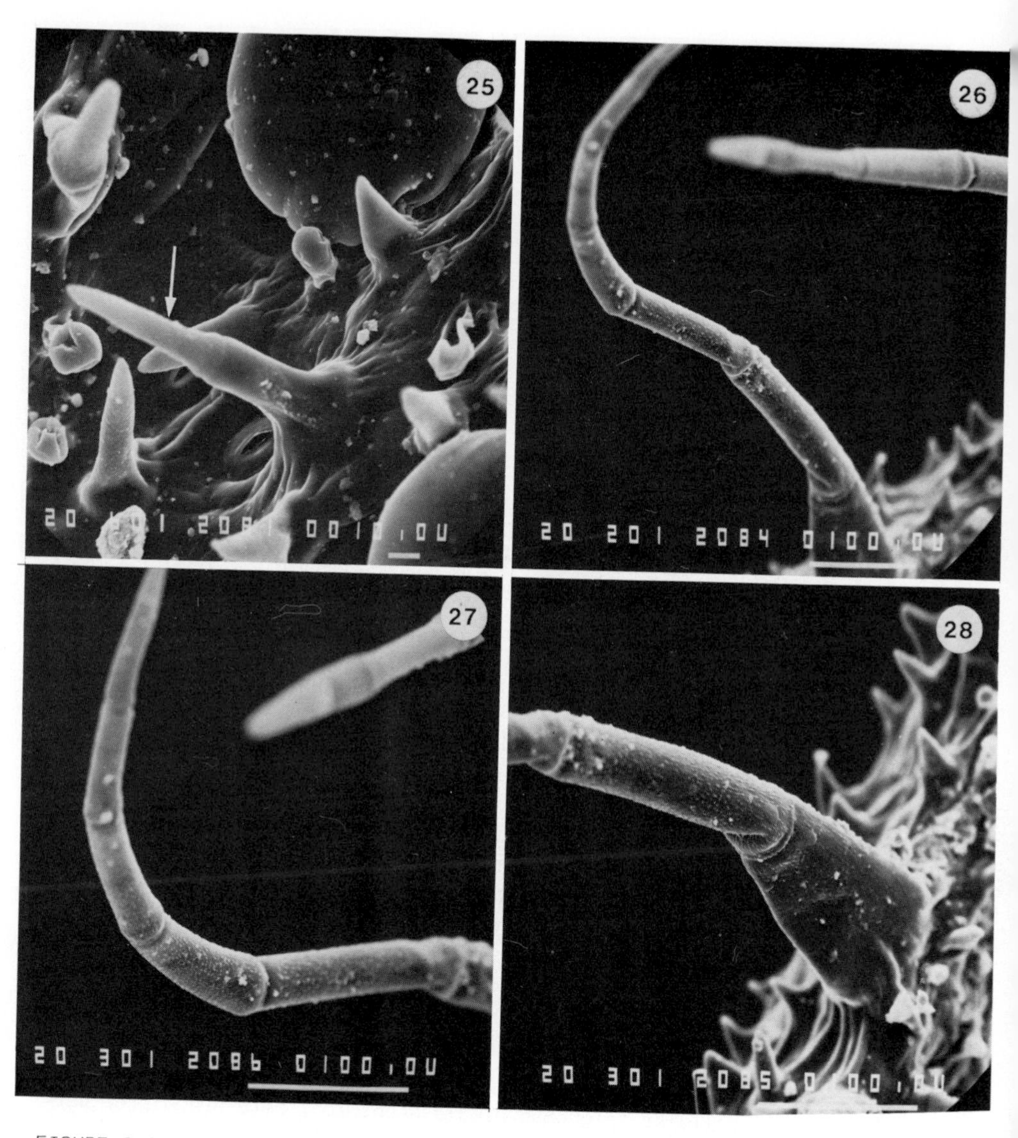

FIGURE 2.25. Isolated two-celled trichome (arrow).

FIGURE 2.26. A marginal multi-celled trichome.

FIGURE 2.27. The upper end of a multi-celled trichome.

FIGURE 2.28. The base of a multi-celled trichome.

The printed number (1000, 100, 10) in the right bottom-corner of each figure represents the scale in µm.

INULA HAIRS - STRUCTURE, ULTRASTRUCTURE AND SECRETION

E. WERKER, A. FAHN

1. INTRODUCTION

The genus *Inula* L. includes 200 species in Europe, Asia and Africa, some of which are known as medicinal plants. Among them *I. viscosa* is a very common Mediterranean perennial herb or chamaephyte, with hairy leaves and young stems and a very strong characteristic smell. The leaves and roots are used in folk medicine: in bath fumes for rheumatic diseases; for inhalations to cleanse the nasal passages; as a herb tea for bronchial and lung diseases; the leaves are also placed on wounds (4, 12).

Glandular hairs of many other genera have already been described (7, 22). They were usually found to secrete mainly one type of material. The glandular hairs of *I. viscosa* and their secretion have been investigated (29), too. Data on the unusually large variety of the types of secretory cells of the hair, on their secreted materials, on the ways by which the materials are secreted to the outside, and on the sites of their synthesis in the cytoplasm are given. However, full interpretation of the results is rather difficult.

It was thought that a comparative investigation of the hairs of other *Inula* species, which might have both different and common features with those of *I. viscosa*, might aid in problems such as identification of cell constituents involved in the production of various secretory materials, and the manner of secretion of different materials to the outside.

Two species were selected for investigation. 1. *I. crithmoides*, a perennial herb, odourless, with a glabrous appearance - though our preliminary examination showed that it has some glandular hairs- used as a medicinal plant for the same purposes as *I. viscosa*. 2. *I. graveolens*, an annual herb, hairy, with a strong camphor-like smell. Preliminary results of this comparison and its significance are reported here.

2. MATERIAL AND METHODS

I. viscosa (L.)Ait. and *I. graveolens* (L.)Desf. were collected from

Margaris N, Koedam A, and Vokou D (eds.): Aromatic Plants: Basic and Applied Aspects
 ISBN 90-247-2720-0.
Printed in the Netherlands.

roadsides and newly disturbed soil in Jerusalem, and *I. crithmoides* L. from seashore rocks near Caesarea.

Light microscopy. Fresh material was used for the following histochemical tests; lipids: Sudan IV and Nile blue (13, 28); polysaccharides other than cellulose: ruthenium red, alcian blue (24) and tannic acid-$FeCl_3$ (18); proteins: mercuric bromophenol blue (16), Commassie blue (9) and eosin (14).

Scanning electron microscopy. Fresh material and material dehydrated in graded ethanol series and dried to the critical point, were coated with gold, and examined with a Jeol scanning electron microscope JC M35.

Transmission electron microscopy. Leaf segments of *I. viscosa* and *I. crithmoides* were fixed with 3%-5% glutaraldehyde for 2h and postfixed in 2% osmium tetroxide for 1.5h, both buffered with 0.2M sodium cacodylate at pH 7.2. The fixed material was dehydrated either in ethanol or in acetone at 4°C and embedded in Spurr's resin (25). Sections were stained with uranyl acetate, followed by lead citrate, and examined with a Philips EM300 electron microscope.

3. RESULTS

3.1. Structure

In all three species biseriate glandular hairs and non-glandular, long multi-cellular uniseriate hairs are present on the leaves and young stems. They are densely arranged, especially on young leaves in *I. viscosa* and *I. graveolens*, and are sparse in *I. crithmoides* (Figs. 1-3). In the first two species stalked and sessile glandular hairs can be distinguished. Both types consist of a head of 5 (or 4) pairs of secretory cells in *I. viscosa* and 6 (or 5) in *I. graveolens*. The stalk consists of elongated cells, especially in *I. viscosa* (Figs. 1, 4) and highly vacuolated while the cells of the head are flat,with dense cytoplasm. In *I. crithmoides* short and longer hairs can be distinguished (Figs. 3, 5); there is however, no sharp delimitation between head and stalk. The total number of cells is 5-8. The walls of all hair cells of the three species are not cutinized, but they are covered by a thin cuticle.

The cells of the head in all types of hairs are round at the summit. In *I. graveolens* the head is barrel shaped. The three (*I. viscosa*) or four (*I. graveolens*) lowermost cells of the head, the "photosynthesizing cells" (29), contain green chloroplasts. In *I. crithmoides* the colour of the plastids is very faint. Electron micrographs show well developed chloroplasts in *I. viscosa* (Figs. 27, 29).

3.2. Secreted Materials

Histochemical stainings of material in and on the hairs' head are summarized in Table 1.

Table 1. Histochemical identification of secreted materials of *Inula* species.

Stain	Compound	Colour	Species		
			Inula viscosa	*Inula graveolens*	*Inula crithmoides*
Sudan IV	Lipids	Red	+	+	+
Nile blue	Neutral lipids	Pink	+	?*	-
	Acidic lipids & phospholipids	Blue	+	+	+
Ruthenium red	Pectins & hemicellulose	Pink	+	+	+
Tannic acid-$FeCl_3$	Pectins	Brown-black	+	+	+
Alcian blue	Acidic (pectins) or sulphated polysaccharides	Blue	+	+	+
Commassie blue	Proteins	Blue	+	±	±
Bromophenol blue-HgCl	Proteins	Blue	+	±	±
Eosin	Proteins	Pink	+	±	±

± staining inside cells only, not on hair

* very faint, inside cells only

The glandular hairs of *I. viscosa* secrete polysaccharides, at least partly pectinaceous, lipids, both neutral and acidic (or phospholipids), and proteins. *I. graveolens* and *I. crithmoides* hairs also secrete polysaccharides, and acidic lipids (or phospholipids). They secrete no or only very small amounts of proteinaceous materials. It remains uncertain whether *I. graveolens* secretes neutral lipids. Other substances, for which no tests were made, might also be present, since the secreted drops of *I. viscosa* and *I. graveolens* become yellowish-brown at later stages of leaf development.

I. viscosa hairs secrete the greatest amount of materials, starting from the time the leaf is ca 2 mm long up to its senescence. The amount of secreted materials on each hair increases rapidly until it overflows, nearly covering the hair. A stain on the ground under the plant can sometimes be seen. In very young leaves where hairs are very dense the leaf is covered with the secreted material. In *I. crithmoides* the drops seen upon the head on young leaves do not increase much in size with age and they remain at the site of secretion. The amount of secreted materials in *I. graveolens* is somewhat lower than that of *I. viscosa*.

3.3. Secretion

The different materials produced in the various head cells at various stages during leaf development all have to be secreted through the outer wall and cuticle. In principle the three species have a similar pattern of secretion, consisting of the following processes in the summit cells:

1. Drops secretion through tiny pores in the cuticle (Fig. 6).
2. Accumulation of additional secretory material between the upper cell wall and cuticle of each or both summit cells (Figs. 8, 9).
3. Bursting of the cuticle (Figs. 11, 12, 14).
4. Secretion of other materials at the apex between the summit cells (Fig. 10) and/or as drops throughout the outer cell wall (Fig. 13). Stages 1 and 2 partly overlap (Fig. 7).

Secretion through the walls of the rest of the head cells takes place through pores at junctions between cells (Figs. 5, 15, 16).

The outer walls of the summit cells of *I. viscosa* and *I. crithmoides* were found to be porous when examined with the electron microscope. In *I. viscosa* there is also an inner wall layer with protrusions into the cell (Fig. 18), which were absent in the young but already secreting hairs of

I. crithmoides examined (Fig. 19). The walls of all the other head cells are traversed by a network of canals (Figs. 20, 21), especially well developed at the zones where the material is ultimately secreted (Fig. 15). With the light microscope the outer walls of the summit cells, except close to their bases, were entirely stained the way the secreted materials were, while in the rest of the cells staining occurred only in a fine line in the middle of the cell walls, and in spots at certain corners, in and on the cells.

3.4. Site of synthesis of the secretory materials

Hairs of leaves of *I. viscosa* at various stages of development have been examined with the electron microscope (29). Preliminary comparative work on *I. crithmoides* shows that in both species cell pairs located at different heights, i.e. the summit cells, the pair below them and the photosynthesizing cells, have different sequences of ultrastructural changes. There is also evidence for the involvement of different cell constituents in secretion: at each stage a different one was found in great abundance in the cell, either producing vesicles or accumulating osmiophilic droplets, which ultimately reach the plasmalemma. The following cell constituents are involved: (a) Endoplasmic reticulum (ER). Smooth ER, tubular and cisternal, was found to be very active at various phases of all cell types (Fig. 17). In the photosynthesizing cells of *I. viscosa* it was found throughout most stages of leaf development (Fig. 18). Rough ER was also present but in much smaller amounts. (b) Plastids. Leucoplasts were found in the summit cells in both species, accumulating osmiophilic material, each in its own manner (Figs. 24-26). In the cell pair below summit cells in *I. viscosa* there were plastids with well developed grana but retaining their prolamellar bodies. These plastids and the chloroplasts of the photosynthesizing cells below them accumulated osmiophilic material as droplets both in the frets and grana, as well as in peripheral reticulum (Fig. 28). (c) Golgi bodies were found to be very active at some stages in both species (Figs. 24, 29). (d) Mitochondria in *I. viscosa* were found either intact, accumulating osmiophilic droplets (Fig. 23), or disrupted, discharging, content into vesicles (Fig. 22).

4. DISCUSSION

Morphologically the glandular hairs of the three *Inula* species are basically similar. Comparison between them suggests that a sharp distinction between a secreting head and a non-secreting stalk cannot be made. Moreover, in *I. viscosa* neutral lipids were found in stalk cells (29) and protrusion of a drop of secreted material has been observed through one of the upper stalk cells (Fig. 16).

Tubular ER was suggested to be involved in terpenoids, flavonoids and fats biosynthesis (2, 10, 11, 21, 32). This is the cell constituent most common in *I. viscosa* hairs and the first to appear prior to the onset of secretion. Plastids were also considered to be involved in the production of lipophilic substances (1, 6). Three types of plastids are involved in the secretion in *I. viscosa*. In *I. crithmoides*, one additional manner of accumulation of osmiophilic material in plastids has been observed so far. Golgi bodies were reported to secrete both acid and neutral polysaccharides. These organelles were found in abundance at certain stages in both *Inula* species. Mitochondria were suggested to be involved, in hairs and other secretory tissues, in both lipid (8, 27) and polysaccharide (3, 30) production. Mitochondria apparently take part in secretion in *I. viscosa* in more than one way. Rough ER, found in *I. viscosa*, is known to be involved in protein production (10, 31). After a thorough ultrastructural study of the hairs of the two other species and a more precise chemical analysis of the secreted materials, the comparison of ways of production of common substances might be worthwhile.

A correlation between the amount of materials produced by the hairs in each species and the chlorophyll of their photosynthesizing cells is evident. This indicates that at least some substances are synthesized *in situ*.

In most glands, materials are secreted either directly through pores in the cuticle, e.g. salt solutions, nectar and lipids (7, 19); or, as some lipids, they are first accumulated under the cuticle and then released through the pores (2); or, after being accumulated, the cuticle eventually bursts and the lipids (5) or mucilages (20, 26) are released. The *Inula* species examined are remarkable in that materials are secreted in all these different ways. The manner of secretion probably corresponds with the characteristics of the secreted materials, e.g. viscosity.

The hairs of *Inula* lack barrier cells with cutinized walls, which are common to many glandular hairs. The presence of a canal network traversing the walls of most head cells renders improbable backflow of the secreted materials. It is not yet clear whether the presence of wall protrusions in the summit cells of *I. viscosa* and their absence in *I. crithmoides* is related to the quantity or quality of the secreted materials (cf. 15).

What does the plant benefit from its secretory materials and how can mankind profit from them? *I. viscosa* plants are not eaten by sheep and goats, apparently due to the smell or taste of its lipid substances, although the presence of druse crystals in most hair and leaf cells (29) might also be a contributing factor. Dried leaves are used to repel insects (4). Small animals are trapped on the plant by the viscid secretion, which might be attributed to the polysaccharides. This plant is said to be used in Spain as fly paper (4). We have, however, no explanation for the secretion of proteins, and this needs investigation of their properties.

Antifungal sesquiterpenic acid with a selective antidermatophitic activity has been isolated from the leaves of *I. viscosa* (23). Further investigation of other compounds with therapeutic qualities might be worthwhile.

Both leaves and roots of *Inula* species are used for therapy. Since roots do not bear grandular hairs, a different secretory tissue can be expected. Secretory ducts were reported in subterranean parts of *Inula* species (17). Preliminary examination of leaves and stems showed that there are narrow secretory ducts on the outer side of phloem strands in *I. crithmoides*, which have sparse glandular hairs, whereas they were not found in *I. viscosa* and *I. graveolens*. It is plausible, therefore, to assume that the active therapeutic materials found in *I. viscosa* are indeed those of the secretory hairs.

REFERENCES

1. Akers CP, Weybrew JA, Long RC. 1978. *Am. J. Bot.* 65:282.
2. Amelunxen F. 1965. *Planta Med.* 13:457.
3. Barlow PW, Sargent JA. 1975. *Protoplasma* 83:351.
4. Dafni A. 1980. The mandrakes give forth fragrance (in Hebrew). Haifa, Gestlit.
5. Dell B, McComb AJ. 1975. *Aust. J. Bot.* 23:373.
6. Dell B, McComb AJ. 1977. *Protoplasma* 92:71.
7. Fahn A. 1979. Secretory tissues in plants. London, Academic Press.
8. Fahn A, Benayoun J. 1976. *Ann. Bot.* 40:857.
9. Fairbanks G, Steck TL, Wallach DFH. 1971. *Biochem.* 10:2606.
10. Gunning BES, Steer MW. 1975. Ultrastructure and the biology of plant cells. London, Edward Arnold.
11. Heinrich G. 1973. *Planta Med.* 23:154.
12. Hochberg M. 1980. Guidance to medicinal plants (in Hebrew). Tel-Aviv, Or-Teva.
13. Jensen EA. 1962. Botanical histochemistry. San Francisco, W. H. Freeman & Co.
14. Johansen DA. 1940. Plant microtechnique. New York, McGraw-Hill.

15. Lüttge U. 1971. Structure and function of plant glands. *Ann. Rev. Plant Physiol.* 22:23.
16. Mazia D, Brewer PA, Alfert M. 1953. *Biol. Bull.* 104:57.
17. Metcalfe CR, Chalk L. 1950. Anatomy of the dicotyledons. Vol. II. Oxford, Clarendon Press.
18. Pizzolato TD. 1977. *Bull. Torrey Bot. Club* 104:277.
19. Schnepf E. 1963. *Flora* 153:1.
20. Schnepf E. 1966. *Ber. Dtsch. Bot. Ges.* 78:478.
21. Schnepf E. 1972. *Biochem. Physiol. Pflanzen* 163:113.
22. Schnepf E. 1974. Gland cells. In: Dynamic aspects of plant ultrastructure, Robards AW (ed.) p. 331. London, McGraw-Hill.
23. Schtacher G, Kashman Y. 1970. *J. Med. Chem.* 13:1221.
24. Scott JE, Quintarelli G, Dellovo MC. 1964. *Histochemie* 4:73.
25. Spurr AR. 1969. *J. Ultrastructure Res.* 26:31.
26. Unzelman JM, Healey PL. 1974. *Protoplasma* 80:285.
27. Vassilyev AE. 1970. *Rastit. Resursy* 6:29.
28. Werker E, Fahn A. 1968. *Nature* 218:388.
29. Werker E, Fahn A. 1981. *Bot. Gaz.* 142:461.
30. Werker E, Kislev M. 1978. *Ann. Bot.* 42:809.
31. Werker E, Vaughan JG. 1976. *Israel J. Bot.* 25:140.
32. Wollenweber E, Schnepf E. 1970. *Z. Pflanzenphysiol.* 62:216.

EXPLANATIONS TO FIGURES

Abbreviations: Ch-chloroplast; ER-endoplasmic reticulum; G-Golgi body; H-head; IW-inner wall layer; M-mitochondrion; NG-non-grandular hair; OW-outer wall layer; P-plastid; Pr-wall protrusion; S-secreted material; R-peripheral reticulum; SC-summit cell; St-stalk; V-vesicle; W-cell wall.

FIGURES 1-3, 6-8. Scanning electron micrographs. FIGURES 1-3. Young leaves; 1. *I. viscosa*; 2. *I. graveolens*; 3. *I. crithmoides*. FIGURE 4. Stalked hair of *I. viscosa* stained with bromophenol blue-HgCl. FIGURE 5. Short hair of *I. crithmoides*. FIGURES 6-8. Summit cells of *I. viscosa*; 6, with drops of secreted material; 7, with drops of secreted material on the elevated portion of cuticle; 8, with elevated cuticle above each cell.

FIGURES 9-15. Scanning electron micrographs. FIGURE 9. *I. crithmoides*; cuticle is elevated above both summit cells. FIGURE 10. *I. viscosa*, secreted material protruding from summit cells. FIGURES 11, 12. Cuticle on summit cells torn; 11, *I. viscosa*; 12, *I. crithmoides*. FIGURE 13. *I. graveolens*; drops of material protruding through summit cell wall. FIGURE 14. *I. crithmoides*; cuticle on each summit cell is torn. FIGURE 15. *I. viscosa*; drop of secreted material between a photosynthesizing cell and a cell below summit cell.

FIGURE 16. Scanning electron micrograph. *I. viscosa*; drop of material protruding from a stalk cell. FIGURES 17-22. Electron micrographs. FIGURE 17. Summit cell of *I. viscosa*; inner wall layer with protrusions, tubular ER with moderately electron-dense material. FIGURE 18. ER in a photosynthesizing cell of *I. viscosa*. FIGURE 19. *I. crithmoides*; walls of summit cells are devoid of protrusions. FIGURES 20, 21. Walls of cells below summit cells with canal networks; 20, *I. viscosa*; 21, *I. crithmoides*. FIGURE 22. Summit cell of *I. viscosa*, mitochondria connected with vesicles.

FIGURES 23-29. Electron micrographs. FIGURES 23-25. Summit cells of *I. viscosa*; 23, mitochondria with osmiophilic droplets; 24, plastid with large osmiophilic drop and membranes, Golgi bodies and many vesicles are present; 25, plastid at a later stage of development filled with osmiophilic material. FIGURE 26. *I. crithmoides*; plastid is filled with small osmiophilic droplets. FIGURE 27. *I. viscosa*; chloroplastid of a photosynthesizing cell with peripheral reticulum. FIGURE 28. The same, but at a later stage of development, with osmiophilic droplets and part of peripheral reticulum. FIGURE 29. Cell at the middle of head, with a plastid, Golgi bodies and ER.

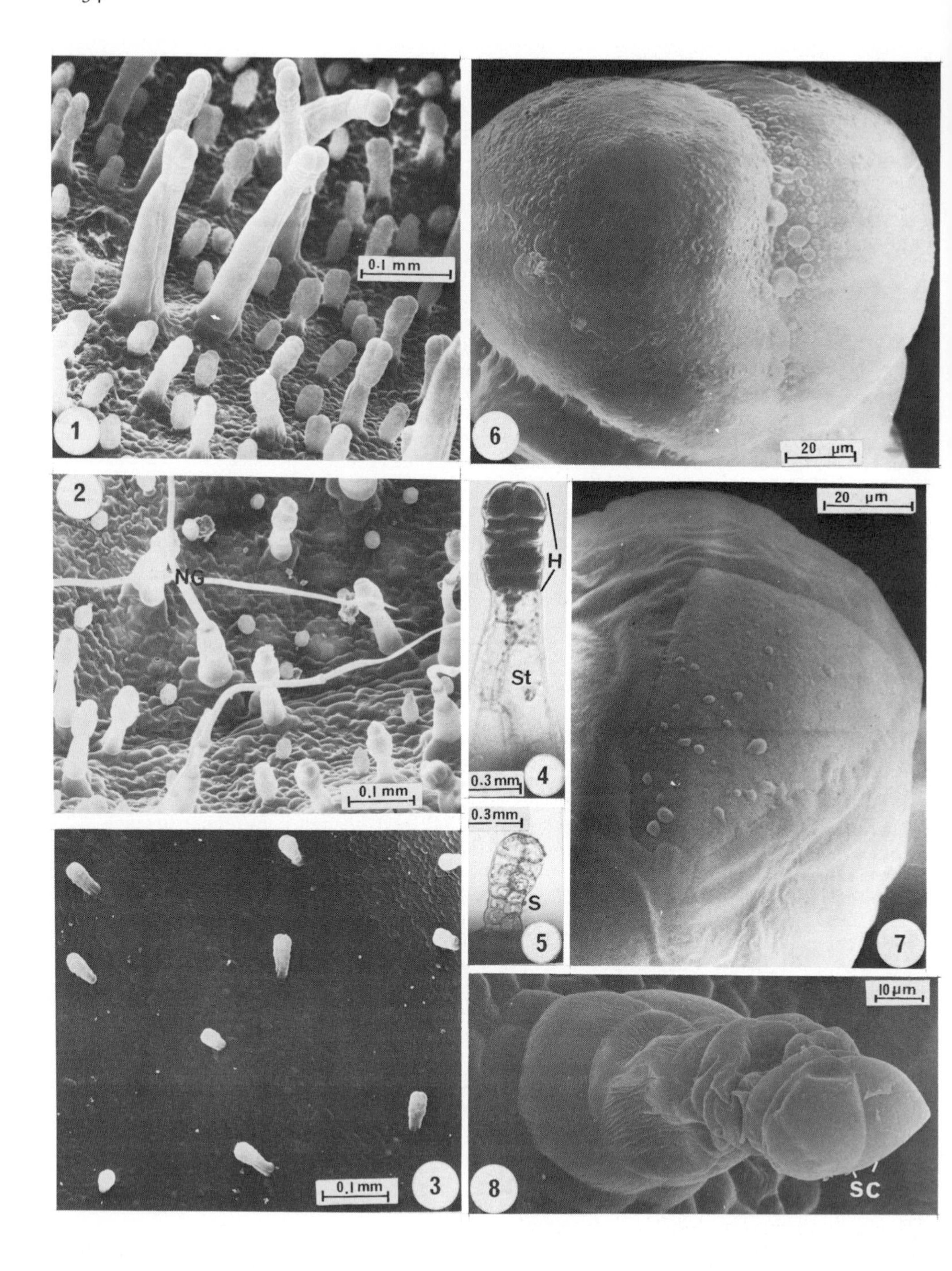
0.1 mm
1
6
20 μm
2
NG
0,1 mm
H
St
0.3 mm
4
20 μm
0.3 mm
S
5
7
10 μm
0,1 mm
3
8
SC

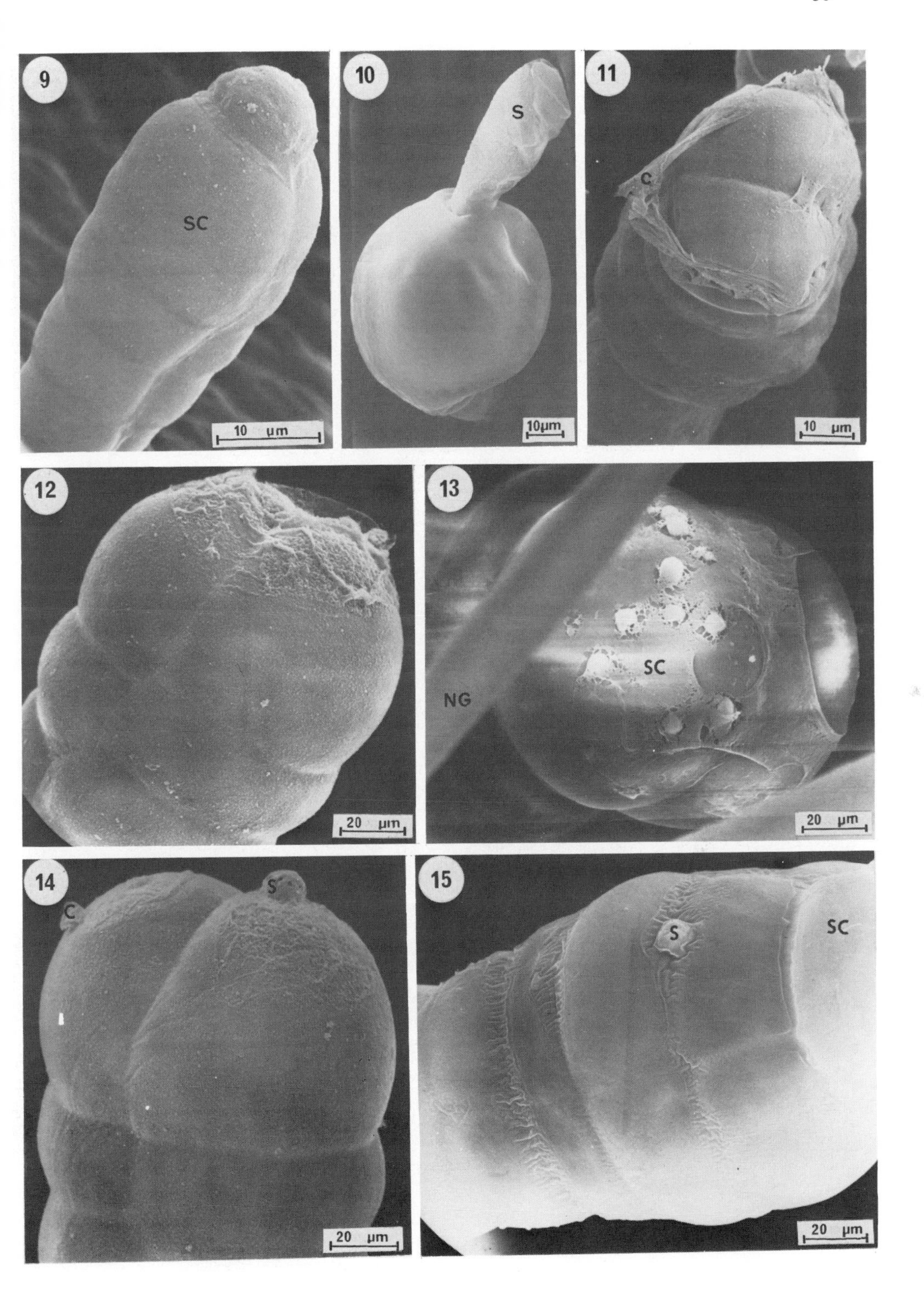
9
SC
10 μm
10
S
10μm
11
C
10 μm
12
20 μm
13
SC
NG
20 μm
14
C
S
20 μm
15
S
SC
20 μm

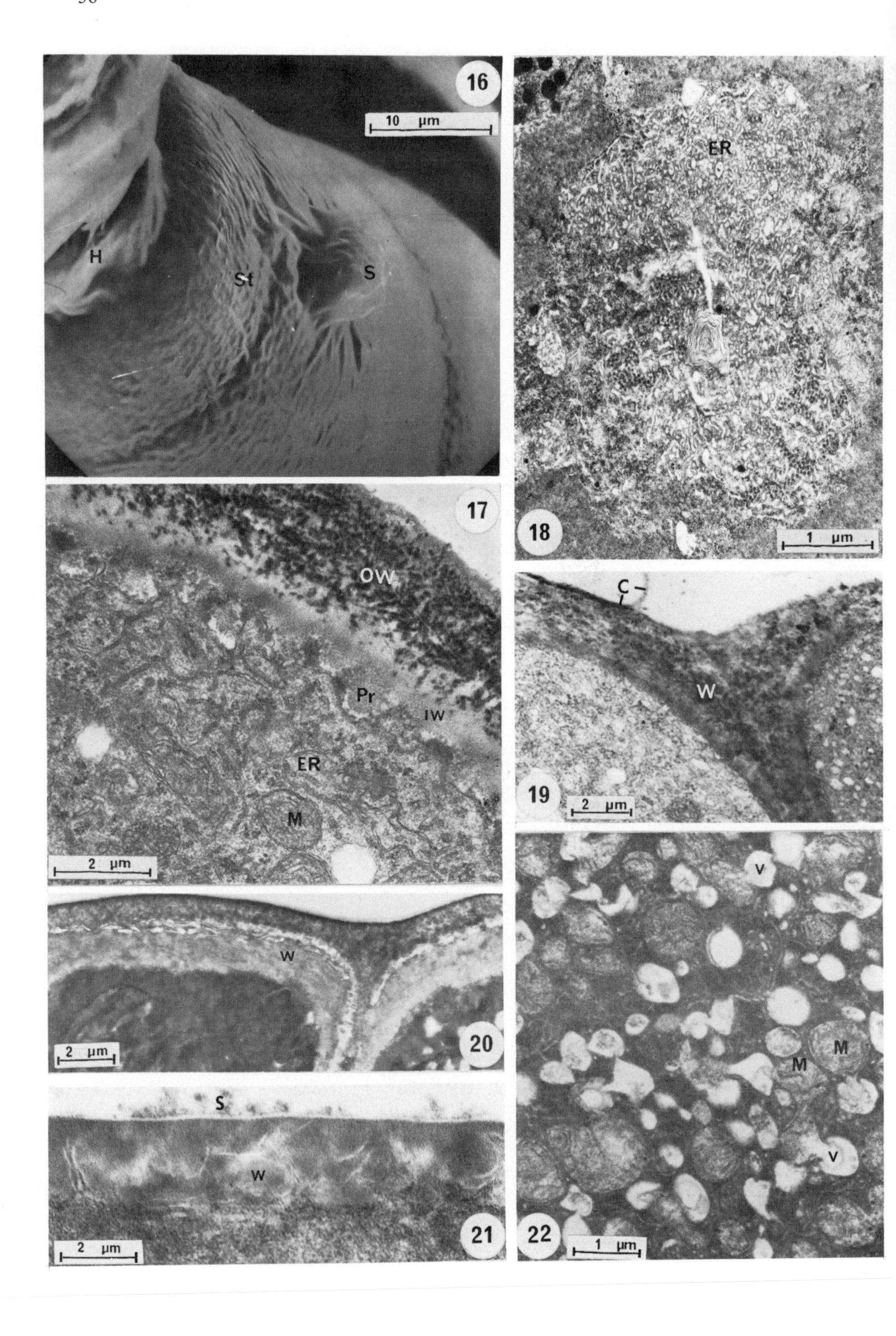
16
10 µm
H
St
S
17
OW
Pr
IW
ER
M
2 µm
18
ER
1 µm
19
C
W
2 µm
20
W
2 µm
21
S
W
2 µm
22
V
M
M
V
1 µm

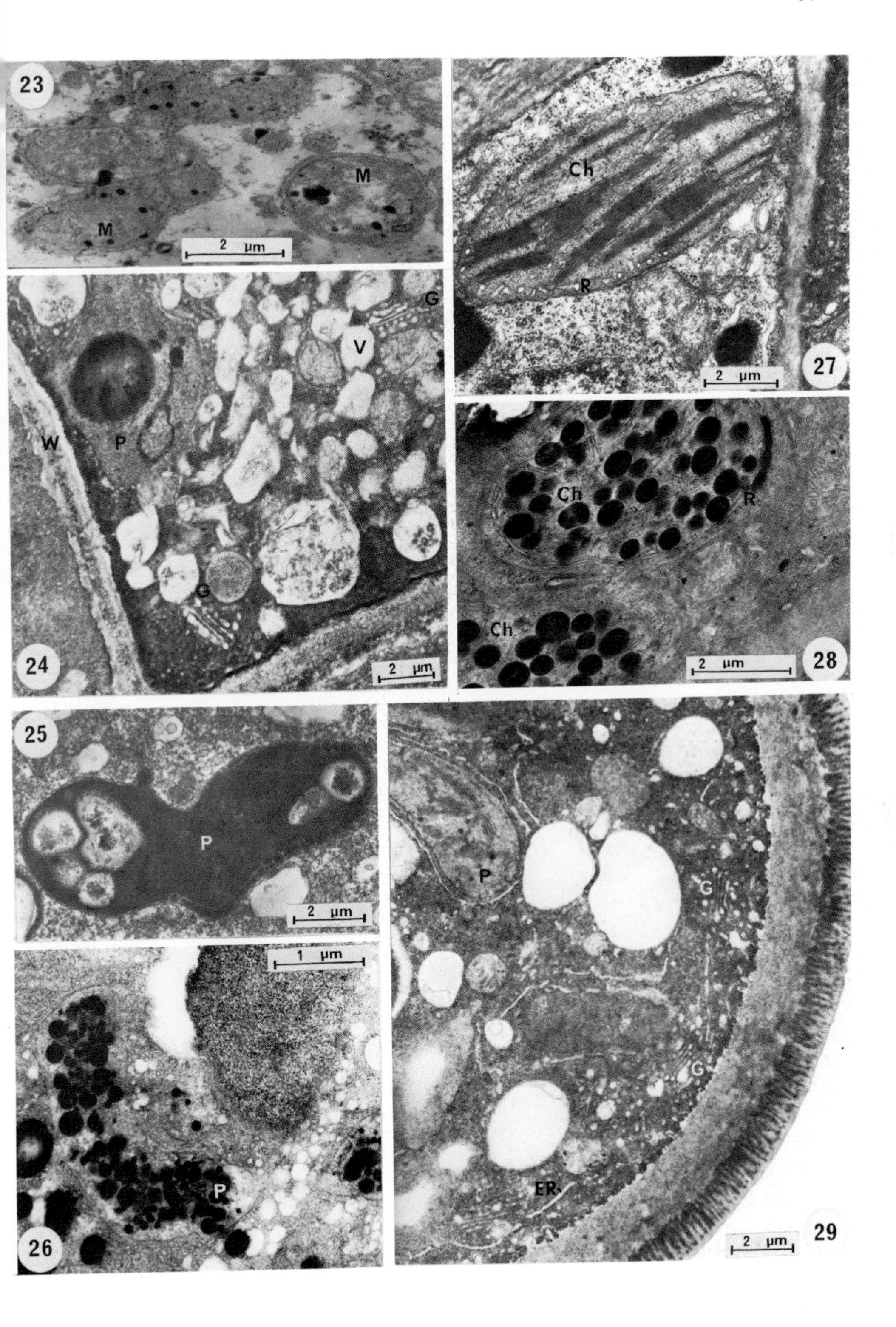
23
M
M
2 µm
24
W
P
V
G
G
2 µm
25
P
2 µm
26
1 µm
P
27
Ch
R
2 µm
28
Ch
R
Ch
2 µm
29
P
G
G
ER
2 µm

POLLEN MORPHOLOGY OF THE GENUS *ORIGANUM* L. AND ALLIED GENERA

S.Z. HUSAIN, V.H. HEYWOOD

1. INTRODUCTION

Over the last thirty years palynological characters have become widely used in flowering plant systematics and, in addition to providing useful comparative data, have at times provided indications of relationships for groups of otherwise uncertain affinity (1, 2, 12). Angiosperm families in general can be divided arbitrarily into two types based on the pollen features: some angiosperm families are stenopalynous, with mainly uniform pollen morphology; the others are eurypalynous, in that they are characterized by a great array of pollen types differing in size, shape, aperture form and exine stratification.

The Labiatae are a stenopalynous family (3). Two main pollen types occur: three-or four-colpate grains, which are often regarded as less specialized, and six-colpate grains (9). Other types of grains are uncommon but 8-colpate grains have been reported (10, 11).

2. PROCEDURE

2.1. Material and Methods

The pollen samples investigated were obtained exclusively from herbarium specimens.* At least three pollen samples from thirty species of *Origanum*, three species of *Thymus*, three species of *Satureja*, two species of *Micromeria*, one species with two subspecies of *Hyssopus*, one species of *Melissa*, one species of *Lycopus*, and three species of *Calamintha* were examined, using light and scanning electron microscope (SEM) techniques. All pollen samples were acetolysed using the acetolysis technique modified from Erdtman (4) prior to light and scanning electron microscopy.

2.1.1. Acetolysis method. Anthers of flowers from herbarium material were dissected. The anthers were softened and crushed in 2-5 drops of

* see Appendix 1

Margaris N, Koedam A, and Vokou D (eds.): Aromatic Plants: Basic and Applied Aspects
© 1982. Martinus Nijhoff Publishers, The Hague/Boston/London. ISBN 90-247-2720-0.
Printed in the Netherlands.

detergent (Tween 80) for 3-5 minutes. 1-2 drops of lactic acid were added to stop further expansion of the pollen grains. The resulting suspension was centrifuged for 2 minutes at 2000 rpm, and the supernatant decanted. The pollen pellet was re-suspended in glacial acetic acid, centrifuged and the supernatant decanted. The pollen pellet was re-suspended again, but in freshly prepared acetolysis mixture, acetic anhydride: concentrated sulphuric acid (9:1) and heated at 100°C for 2-2.5 minutes. After centrifugation the resulting dark brown-black pollen pellet was washed three times in distilled water. The pollen was then divided to provide material for both light and scanning electron microscopy.

2.1.2. Preparation for scanning electron microscopy. The sample was passed through a brief (3-5 minutes) dehydration series of 30%, 50% and 95% ethanol, and finally transferred into acetone. A suspension of the acetolysed pollen grains in acetone was placed on a clean stub and allowed to dry. The stubs were then coated with 1.3 µm of platinum and examined in a Jeol JSM-35 scanning electron microscope at 15 kv.

2.1.3. Preparation of light microscopy. The remaining acetolysed pollen material in the other tube was mixed thoroughly with 50% glycerol, centrifuged and the supernatant decanted. The tube was immediately inverted, and left to drain, overnight at room temperature, or in an oven at 40°C for one hour. The pollen was mounted in glycerine jelly on a microscope slide and sealed with paraffin wax. The prepared slides were examined and dimensions measured (Table 1,2 and Plate 1), with a Watson micro-system 70 microscope and photographs were taken with a Zeiss photomicroscope. All measurements are based on approximately twenty pollen grains of each species.

3. RESULTS

The terminology of tectal sculpturing patterns used was, wherever possible, that of Hideux and Ferguson (6).

3.1. Pollen description of *Origanum* species and allied genera

Palynological results for 30 species of genus *Origanum* and 14 species from allied genera are summarized in Tables 1 and 2. All pollen grains examined were six-colpate, and oblate, subsphaeroidal or prolate in shape. The polar axis ranged from 25-49 µm, and the equatorial diameter from 21-41 µm. The six colpi are equal in size and meridionally positioned at right angles to the equator. The most useful taxonomic characters are equatorial diameter, polar axis and the sculpturing of the tectum as seen under the scanning electron microscope (SEM). Using these features the pollen grains examined appear, without confirmation provided from sections of the exine, to be suprareticulate or foveolate, and can be broadly divided into three types.

3.1.1. Type 1 (Plates 2 and 3). All the grains in this type basically have a simple, perforate (foveolate) tectum, comprising a primary network of smooth to coarse, thick tectal ridges (muri), surrounding spaces (lumina) within which the tectum has 1-3 small scattered perforations (puncta). The tectum is not clearly reticulate above the level of the puncta. The length of the colpus, expressed as a percentage of the total pollen length ranges from 53-69%. Polar axis measuring (28-)39(-53) μm and equatorial diameter (25-)35(-50) μm. There is a very prominent thickening of the exine at the polar region. The tectal ridges are more or less smooth and in some cases granulated in the polar regions. The following species fall into this type:

O. dubium	*Lycopus europaeus*
O. majorana	*Micromeria fruticosa*
O. onites	*M. nervosa*
O. syriacum	*Satureja montana*
Calamintha grandiflora	*S. obovata*
C. sylvatica	*S. thymbra*
C. ascendens	*Thymus ciliatus*

The size and shape of the groups of puncta exhibit a continuous range of variation within the genera, the only differential character between the species being the number of puncta in each group. The grains range from small to large.

3.1.2. Type 2 (Plates 4 and 5). Tectum reticulate with crowded puncta clearly associated into groups of up to six, surrounded by a poorly defined reticulum of weakly differentiated muri, scarcely raised above the level of the tectum. The length of the colpus, expressed as a percentage of the total pollen length ranges from 65%-69%. Polar axis measures (32-37(-42) μm, and the equatorial diameter measures (31-)35(-41) μm. The following species fall into this type:

O. compactum	*O. ramonense*
O. ehrenbergii	*O. virens*
O. heracleoticum	*O. viride*
O. isthmicum	*O. vulgare*

Table 1. Pollen characters of genus *Origanum* species as observed by light microscopy (LM) and scanning electron microscopy (SEM).

TAXA	Type	Shape*and reticulation	Length in polar axis (P)			Equatorial diameter (E)			Ratio (P/E)
			Max	Mean	Min	Max	Mean	Min	
Origanum dubium	1	Sc,P	(38-)	34	(-26)	(30-)	27	(-22)	1.3
O. majorana	1	Sc,P	(36-)	34	(-25)	(28-)	27	(-21)	1.3
O. onites	1	C,P	(39-)	28	(-28)	(35-)	27	(-24)	1.2
O. syriacum	1	Sc,P	(35-)	34	(-29)	(32-)	29	(-27)	1.2
O. compactum	2	C,R	(42-)	36	(-32)	(41-)	36	(-32)	1.0
O. ehrenbergii	2	C,R	(41-)	36	(-32)	(39-)	34	(-32)	1.1
O. heracleoticum	2	Sc-C,R	(39-)	34	(-31)	(39-)	34	(-35)	1.0
O. isthmicum	2	Sc-C,R	(39-)	36	(-32)	(34-)	33	(-31)	1.1
O. laevigatum	2	C,R	(40-)	37	(-33)	(39-)	36	(-32)	1.0
O. micranthum	2	Sc-C,R	(38-)	36	(-31)	(38-)	35	(-32)	1.0
O. microphyllum	2	C,R	(39-)	32	(-28)	(32-)	31	(-22)	1.1
O. ramonense	2	Sc-C,R	(39-)	36	(-34)	(35-)	33	(-31)	1.1
O. virens	2	C,R	(38-)	34	(-29)	(38-)	33	(-28)	1.0
O. viride	2	C,R	(37-)	34	(-28)	(36-)	34	(-27)	1.1
O. vulgare	2	Sc-C,R	(42-)	37	(-32)	(39-)	36	(-34)	1.0
O. akhdarense	3a	Sc-C,R	(38-)	34	(-32)	(36-)	33	(-28)	1.0
O. hypericifolium	3a	Sc-C,R	(39-)	34	(-34)	(36-)	32	(-29)	1.1
O. leptocladum	3a	Sc-C,R	(40-)	35	(-34)	(36-)	33	(-30)	1.0
O. libanoticum	3a	Sc-C,R	(43-)	38	(-36)	(38-)	38	(-33)	1.0
O. lirium	3a	Sc-C,R	(38-)	36	(-34)	(32-)	34	(-28)	1.1
O. rotundifolium	3a	Sc-C,R	(40-)	36	(-34)	(33-)	33	(-30)	1.1
O. scabrum	3a	Sc-C,R	(41-)	36	(-34)	(33-)	33	(-30)	1.1
O. sipyleum	3a	Sc-C,R	(42-)	36	(-35)	(38-)	33	(-30)	1.1
O. vetteri	3a	Sc-C,R	(40-)	37	(-34)	(36-)	32	(-29)	1.1
O. amanum	3b	O,R	(47-)	41	(-36)	(39-)	34	(-31)	1.2
O. cordifolium	3b	O,R	(43-)	40	(-36)	(36-)	34	(-31)	1.2
O. ciliatum	3b	O,R	(44-)	40	(-35)	(37-)	33	(-32)	1.2
O. dictamnus	3b	O,R	(49-)	41	(-35)	(36-)	34	(-31)	1.2
O. saccatum	3b	O,R	(46-)	41	(-34)	(37-)	34	(-32)	1.2
O. tournefortii	3b	O,R	(49-)	43	(-36)	(42-)	35	(-28)	1.2

*Symbol used: Sc=subcircular; Sr=subrectangular; C=circular; O=oval; P=prolate; R=reticulate

Wall ꞁickness	Apocolpial radius	POLLEN GRAIN SIZE IN μm Ectexine thick at poles	Colpus length;(%) of the total pollen length	No of perfora-tions per lumen
2	5	YES	58-63	1,2,3
2	5	YES	58	1,2,3
2	7	YES	63	1,2,3
2	5	YES	69	1,2,3
3	7	YES	69	1,2,3,5,6
3	7	YES	68	1,2,3,4,5,6
2	6	YES	65	1,2,3,4,5
3	7	YES	67	1,2,3,4,5
3	7	YES	68	2,3,4,5
2	6	NO	66	2,3,4,5
2	5	NO	61	1,2,3,4,5,6
3	7	YES	68	2,3,4,5
3	7	YES	69	1,2,3,4,5
3	6	YES	67	1,2,4,5
3	7	YES	66	1,2,3,4,5
3	5	YES	62	2,3,5,6,7
3	5	YES	64	1,2,4,5,6,7
3	5	YES	65	1,2,4,6,7
3	5	YES	64	1,3,4,5,6,7
3	6	YES	67	1,2,3,4,5,7
3	6	YES	66	1,2,4,5,7
3	5	YES	68	1,2,4,5,7
3	5	YES	61	2,3,4,5,6,7
3	5	YES	60	2,3,5,6,7
3	8	NO	74	2,3,4,5,7,8
3	5	NO	69	2,3,4,5,6
3	6	NO	72	2,3,4,5,6,8
3	8	NO	72	3,4,5,6,7,8
3	5	NO	74	3,4,5,6,8
3	8	NO	75	2,4,5,6,8,9,10

Table 2. Pollen characters of allied genera species as observed by light microscopy (LM) and scanning electron microscopy (SEM).

TAXA	Type	Shape*and reticulation	Length in polar axis (P)			Equatorial diameter (E)			Ratio (P/E)
			Max	Mean	Min	Max	Mean	Min	
Calamintha grandiflora	1	O-Sc,P	(53-)	49	(-48)	(50-)	42	(-39)	1.2
C. sylvatica	1	O,P	(46-)	41	(-38)	(39-)	34	(-31)	1.2
C. ascendens	1	O,P	(40-)	39	(-38)	(39-)	33	(-30)	1.2
Lycopus europaeus	1	O-Sc,P	(34-)	30	(-28)	(28-)	25	(-22)	1.2
Micromeria fruticosa	1	Sc-C,P	(39-)	36	(-32)	(38-)	34	(-30)	1.1
M. nervosa	1	Sc,P	(41-)	38	(-36)	(37-)	35	(-34)	1.2
Satureja montana	1	Sc-C,P	(52-)	50	(-47)	(50-)	48	(-43)	1.0
S. obovata	1	Sc-C,P	(39-)	35	(-31)	(34-)	32	(-28)	1.2
S. thymbra	1	Sc-C,P	(36-)	34	(-28)	(35-)	31	(-28)	1.0
Thymus ciliatus	1	C,P	(34-)	29	(-27)	(31-)	28	(-25)	1.0
Melissa officinalis	2	O,R	(38-)	36	(-34)	(32-)	30	(-28)	1.2
Thymus vulgaris	2	Sr-O,R	(36-)	32	(-28)	(31-)	25	(-21)	1.3
Thymus capitatus	3a	Sr-O,R	(42-)	36	(-32)	(38-)	32	(-28)	1.2
Hyssopus officinalis	3b	Sc-C,R	(35-)	30	(-28)	(32-)	28	(-25)	1.0
H. officinalis ssp. *montanus*	3b	C,R	(35-)	31	(-29)	(32-)	31	(-29)	1.0

*Symbol used: Sc=subcircular; Sr=subrectangular; C=circular; O=oval; P=prolate; R=reticulate.

Wall hickness	Apocolpial radius	POLLEN GRAIN SIZE IN µm Ectexine thick at poles	Colpus length;(%) of the total pollen length	No of perforations per lumen
3	8	NO	69	1,2,3
3	7	NO	69	1,2
3	7	NO	68	1,2
2	5	NO	69	1,2
2	6	YES	61	1,2
2	6	YES	63	1,2,3
2	4	YES	57	1,2,3
2	4	YES	59	1,2,3,4
2	4	YES	57	1,2,3
2	6	YES	59	1,2,3
2	5	NO	68	1,2,3,4,5
2	4	YES	65	1,2,3,4,5
2	5	YES	64	1,2,3,4,5,6
2	5	NO	76	2,3,5,7,8,17,19
2	5	NO	69	4,5,8,16,18

O. laevigatum *Melissa officinalis*
O. micranthum *Thymus vulgaris*
O. microphyllum

All the pollen grains examined in this type 2 show a similar surface sculpture to that of type 1, except that the tectum bears a reticulate sculpture enclosing the groups of puncta. The tectal ridges are thicker and much flatter than the subsidiary ridges separating the puncta.

3.1.3. Type 3. Tectum reticulated with crowded puncta which are clearly associated into groups of nine to eighteen, surrounded by a clearly defined reticulum with strongly differentiated muri, distinctly raised above the level of the tectum. The length of the colpus expressed as a percentage of the total pollen length ranges from 60-76%. The polar axis is (32-)38(-49) µm and the equatorial diameter ranging between (28-)34(-42) µm.

All the species included in this type 3 exhibit a basically reticulate pattern above the tectum. The primary network between the puncta is more complex with thick and somewhat raised tectal subsidiary ridges. The lumina are deep, oval-subcircular in the polar region and oval-subrectangular in the equatorial region. The pollen grains are generally large. Two subtypes can be recognized.

3.1.4. Type 3A (Plate 6). The pollen grains show a pronounced thickening of the ectexine at the apocolpium. Perforations per lumen range from 3-8. The length of the colpus expressed as a percentage of the total pollen length is 60-68%. The polar axis is (32-)36(-43) µm, and equatorial diameter (28-)34(-39) µm. The species included in this type 3A are:

O. akhdarense *O. rotundifolium*
O. hypericifolium *O. scabrum*
O. leptocladum *O. sipyleum*
O. libanoticum *O. vetteri*
O. lirium *Thymus capitatus*

3.1.5. Type 3B (Plate 7). The pollen grains are large, subcircular-oval in shape showing no pronounced thickening of the ectexine at the apocolpium. The length of the colpus expressed as a percentage of the total pollen length ranges from 69-76%. Polar axis is (34-)41(-49) µm, and equatorial diameter, (28-)34(-42) µm.

The species included in this type 3B are:

O. amanum *O. saccatum*
O. cordifolium *O. tournefortii*
O. ciliatum *Hyssopus officinalis*

O. dictamnus *H. officinalis* ssp. *montanus*

The reticulate pattern is similar to type 3A, but the primary network is fine, complex with many subsidiary ridges. Tectal ridges are thin, smooth and flat.

4. DISCUSSION

The present results add further evidence for the palynological homogeneity of the family Labiatae (5). On the basis of pollen morphology the species of the genus *Origanum* and representative species of the allied genera examined here can be divided into three types and two sub-types (Tables 1, 2). Type 1 contains all the species of section Majorana of the genus *Origanum*, with the exception of *O. microphyllum* and *O. micranthum*, which fall in type 2 along with the species of section Origanum.

The recent morphological revision by Ietswaart suggests that *O. microphyllum* and *O. micranthum* could form a new section Chilocalyx, as distinct from section Majorana.

Our pollen morphological results suggest that these two species agree with this proposal (8), but phytochemical evidence (7) supports this view with respect to *O. micranthum* only. It does not provide support for the removal of *O. microphyllum* from section Majorana.

Most of the species belonging to section Amaracus *sensu lato* fall into pollen types 3A and 3B (Tables 1, 2).

This ivestigation has also demonstrated a similar palynological continuum of variations in the sculpturing of the representative species of the allied genera. Although it is difficult to separate species within these types, even on the basis of fine details of the reticulum, the size and shape of the reticulum does exhibit a small degree of variation not only across the allied genera, but also within the genus *Origanum*, sufficient to allow the recognition of 3 types and 2 subtypes. Further work as carried out by Nabli with transmission electron microscopy (10) and information from the stratification of the exine might reveal some other interesting characters.

REFERENCES

1. Davis PH, Heywood VH. 1963. Principles of Angiosperm Taxonomy, p. 187. Edinburgh and London, Oliver and Boyd.
2. Erdtman G. 1952. Pollen morphology and plant taxonomy. Angiosperms. Stockholm, Almqvist and Wiksell.
3. Erdtman G. 1945. *Svensk. Bot. Tidskr.* 39.

4. Erdtman G. 1969. Handbook of Palynology, Morphology, Taxonomy-Ecology. Copenhagen.
5. Henderson DM et al. 1968. *Grana Palynol.* 8:70.
6. Hideux MJ, Ferguson IK. 1976. The stereostructure of the exine and its evolutionary significance in Saxifragaceae. In: The evolutionary significance of the exine, Ferguson IK and Muller J (eds.). London and New York, Academic Press.
7. Husain SZ, Markham K. 1981. *Phytochemistry* 20:1171.
8. Ietswaart JH. 1980. A taxonomic revision of the genus *Origanum* (Labiatae The Hague/Boston/London, Leiden University Press.
9. Leitner J. 1942. *Österr. Bot. Zeitschr.* 91:29.
10. Nabli MA. 1976. In: The evolutionary significance of the exine, Ferguson IK and Muller J (eds.). London and New York, Academic Press.
11. Risch C, 1956. *Willdenowia* 1:617
12. Wodehouse RP. 1935. Pollen grains. Their structure, identification and significance in science and medicine. New York and London.

LEGEND TO PLATES 1-7

Plate 1 : Light microscope photographs A-L, showing pollen shape, exine wall thickness and apocolpial radius in *Origanum* and allied genera.

Plates 2-7: Scanning electron micrographs showing pollen shape and surface sculpturing in *Origanum* and allied genera.

Plate 2 : A,B. *Origanum syriacum*
C,D. *O. dubium*
E,F. *Thymus ciliatus*

Plate 3 : A,B. *Calamintha grandiflora*
C,D. *Lycopus europaeus*
E,F. *Micromeria fruticosa*

Plate 4 : A,B. *Origanum compactum*
C,D. *O. microphyllum*
E,F. *Thymus vulgaris*

Plate 5 : A,B. *Origanum ramonense*
C,D. *O. virens*
E,F. *Melissa officinalis*

Plate 6 : A,B,C,D. *Origanum scabrum*
E,F. *Thymus capitatus*

Plate 7 : A,B. *Origanum dictamnus*
C,D. *O. tournefortii*
E,F. *Hyssopus officinalis*

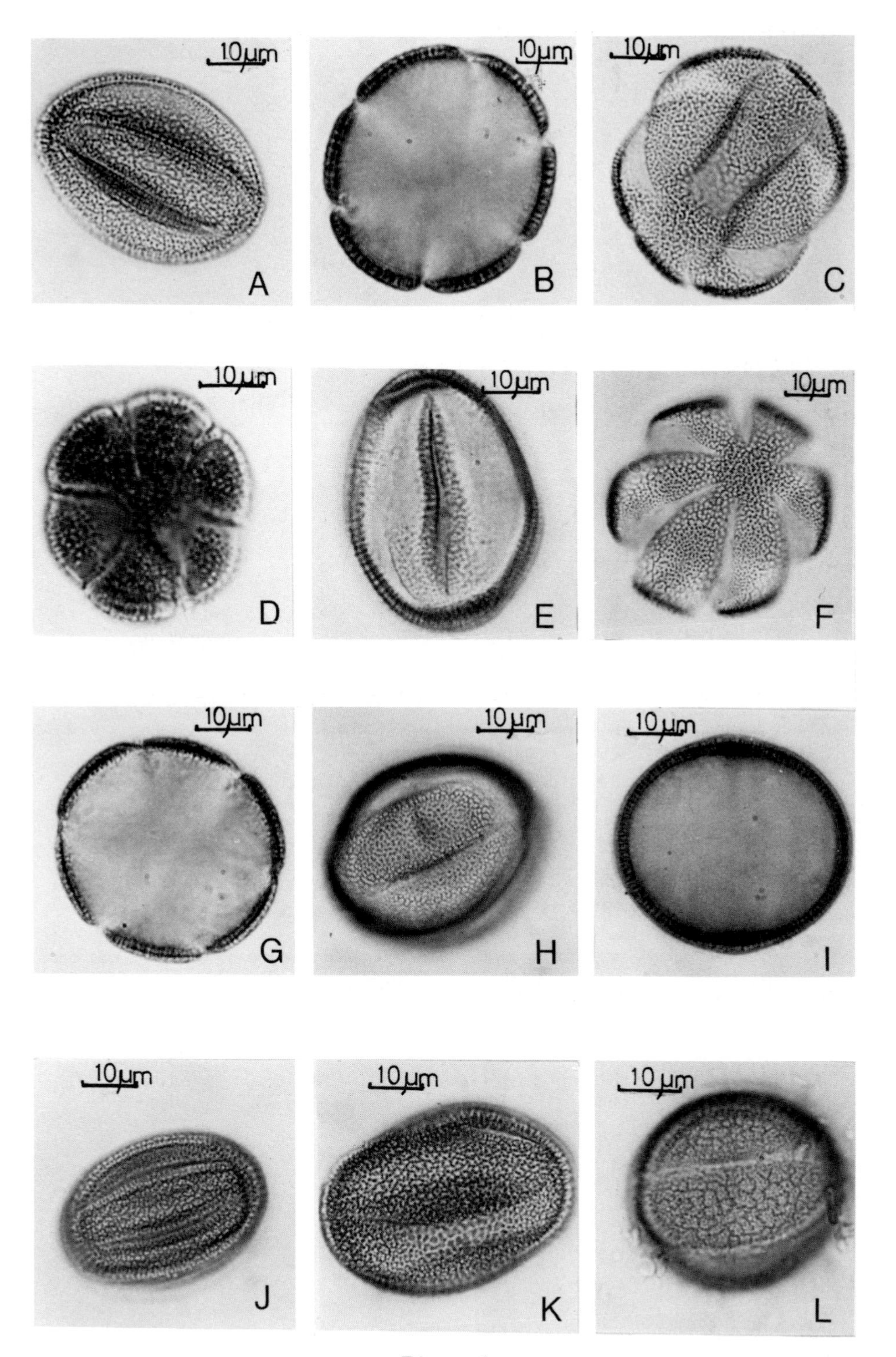
10µm
A
10µm
B
10µm
C
10µm
D
10µm
E
10µm
F
10µm
G
10µm
H
10µm
I
10µm
J
10µm
K
10µm
L

Plate 1

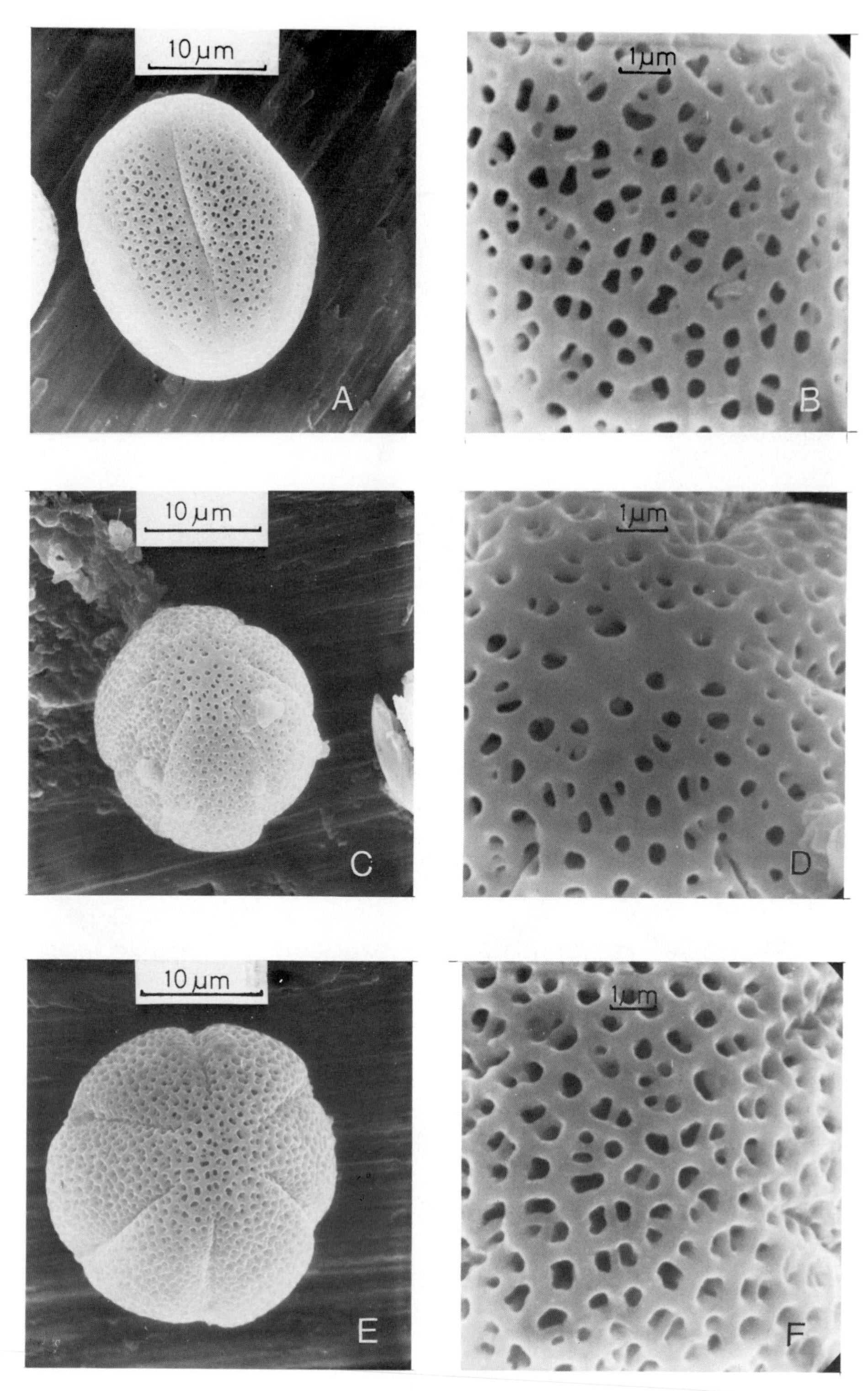

Plate 2

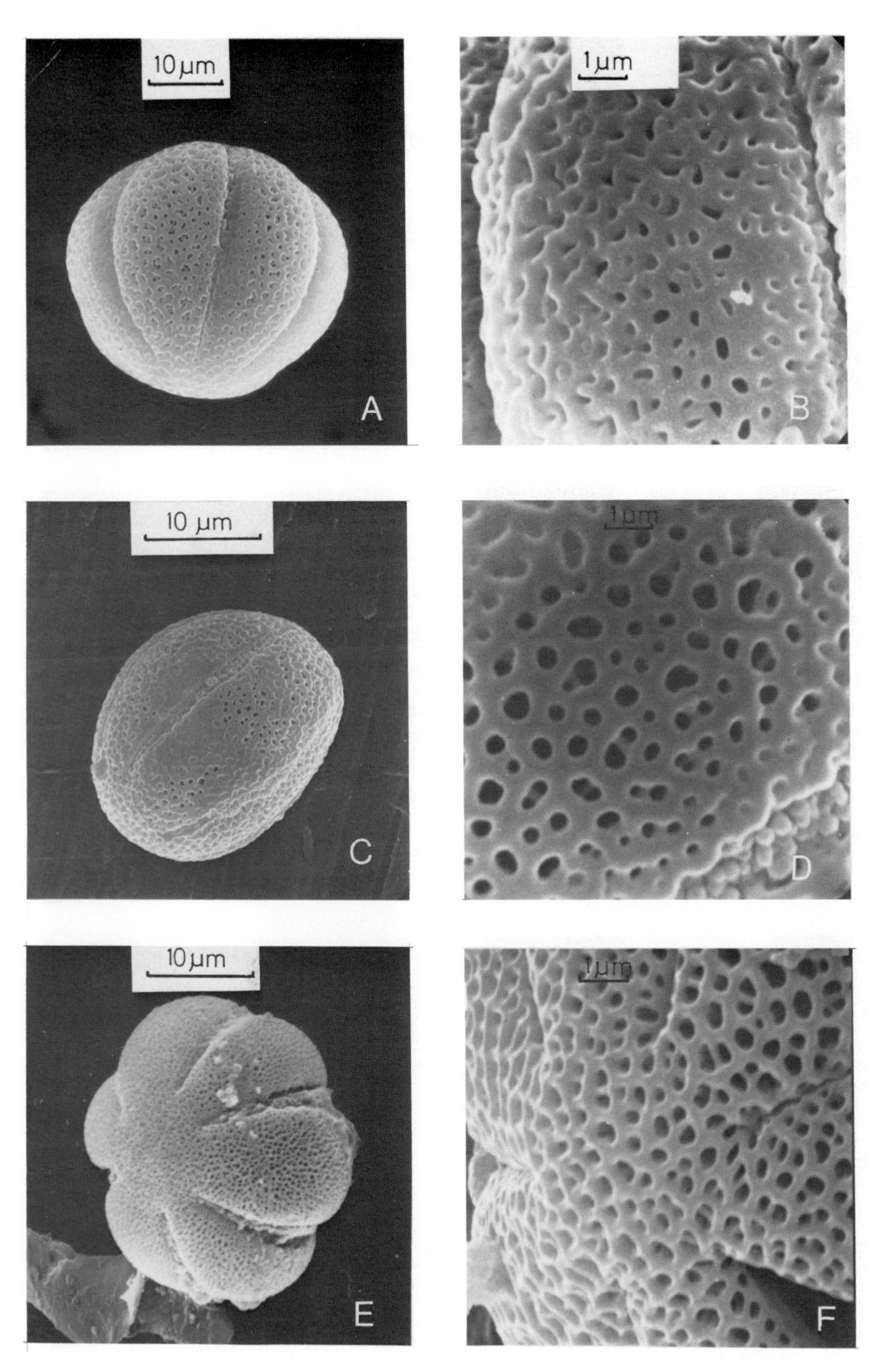

Plate 3

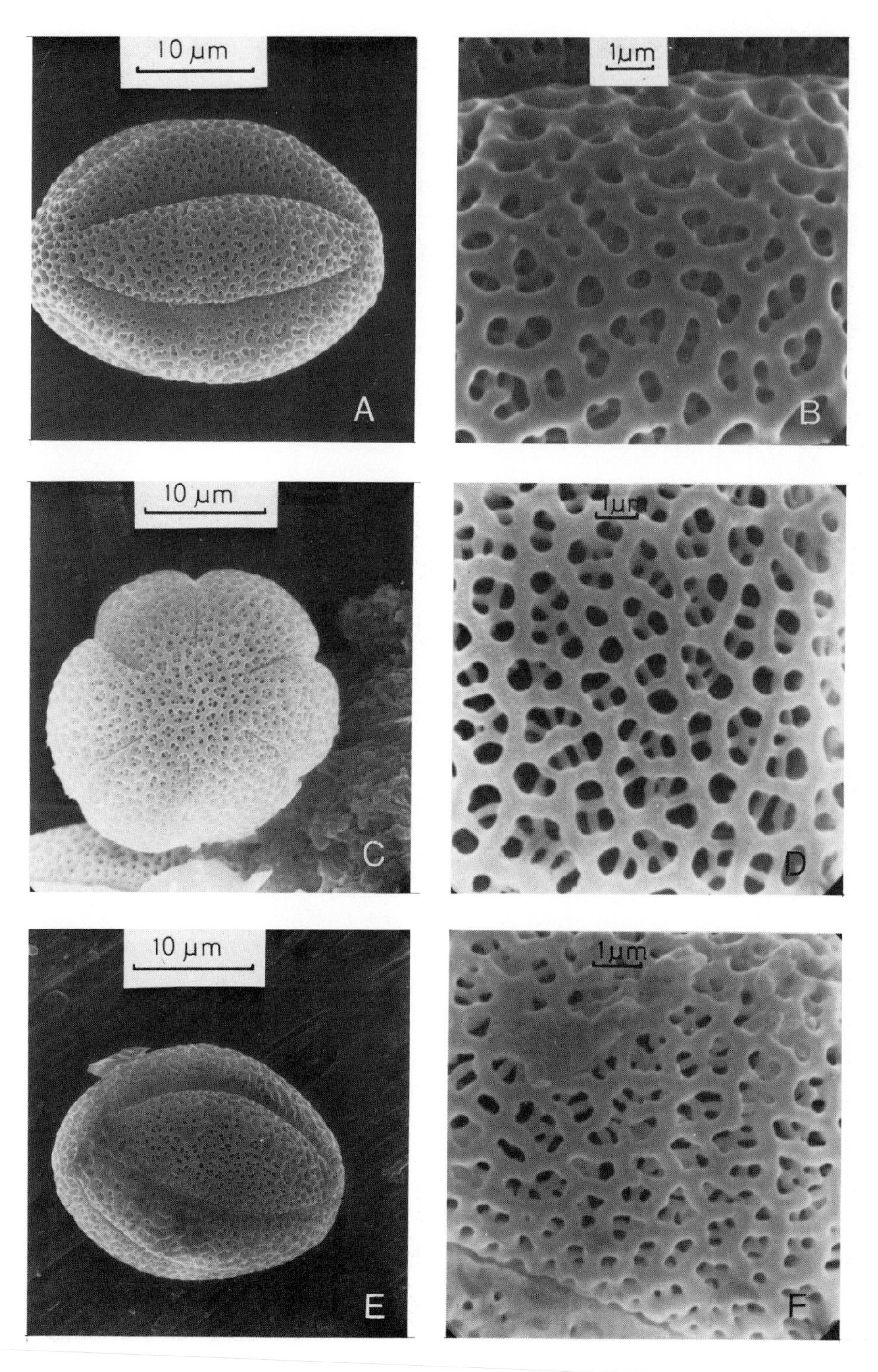

Plate 4

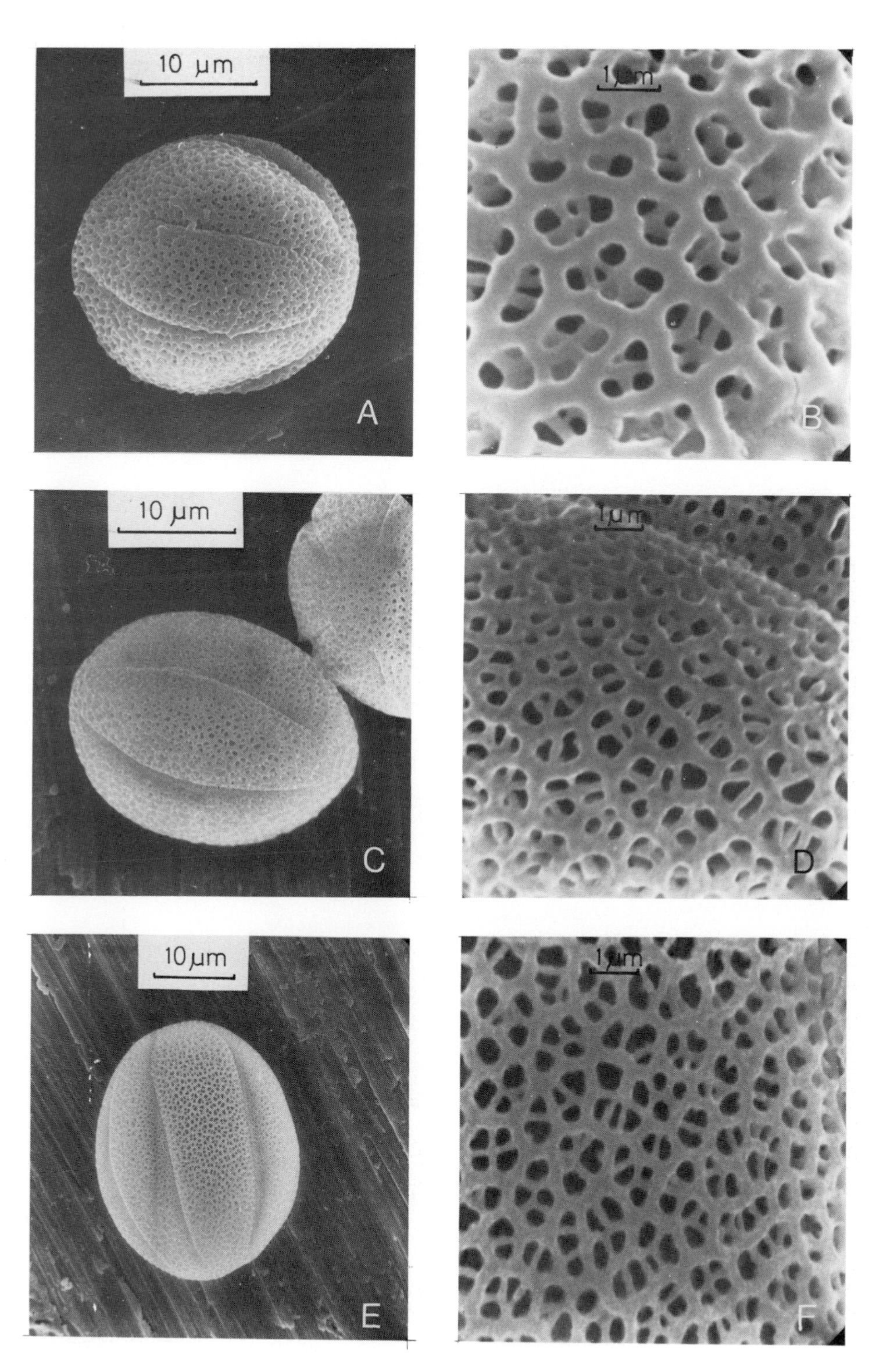

Plate 5

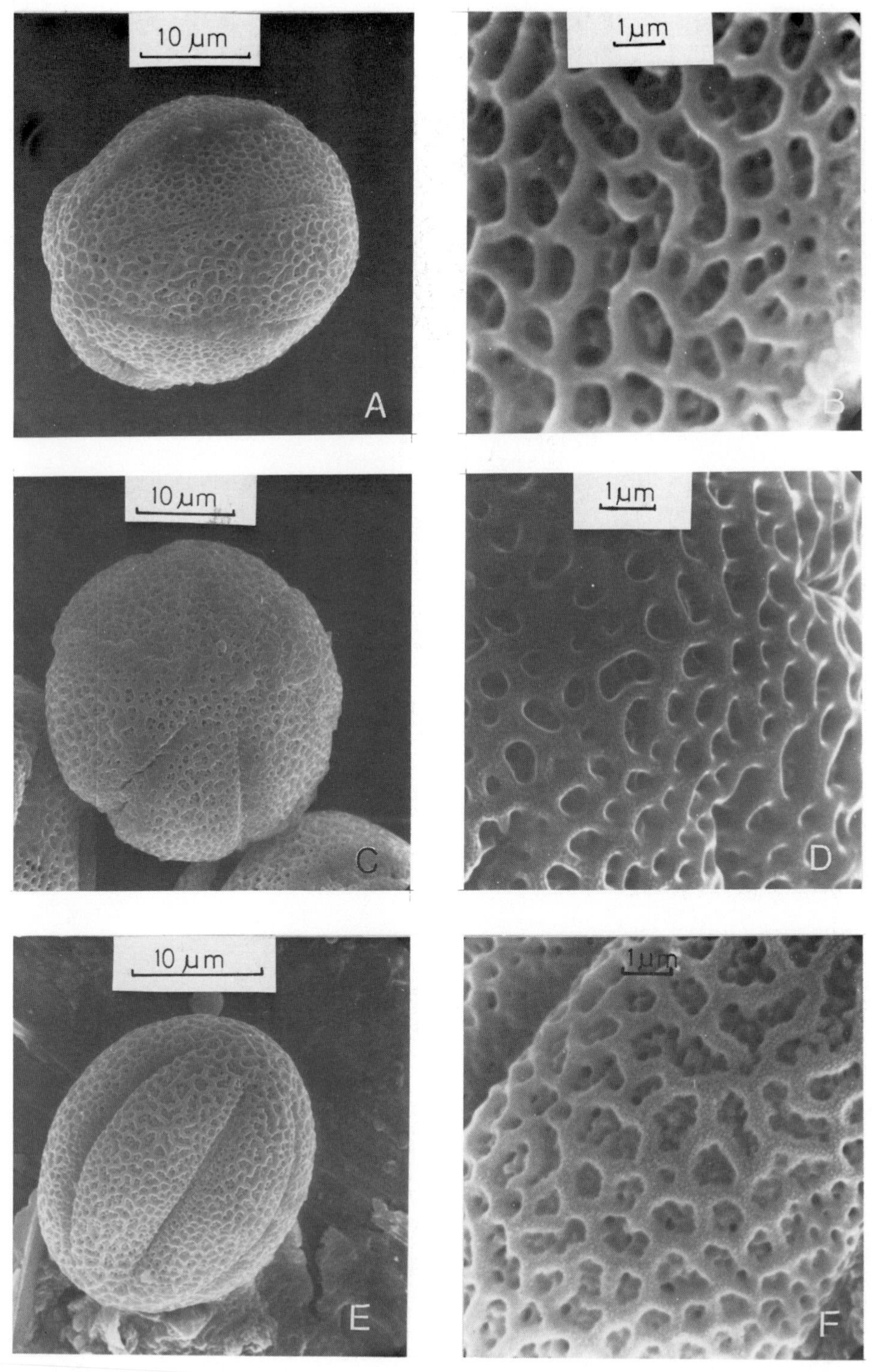

Plate 6

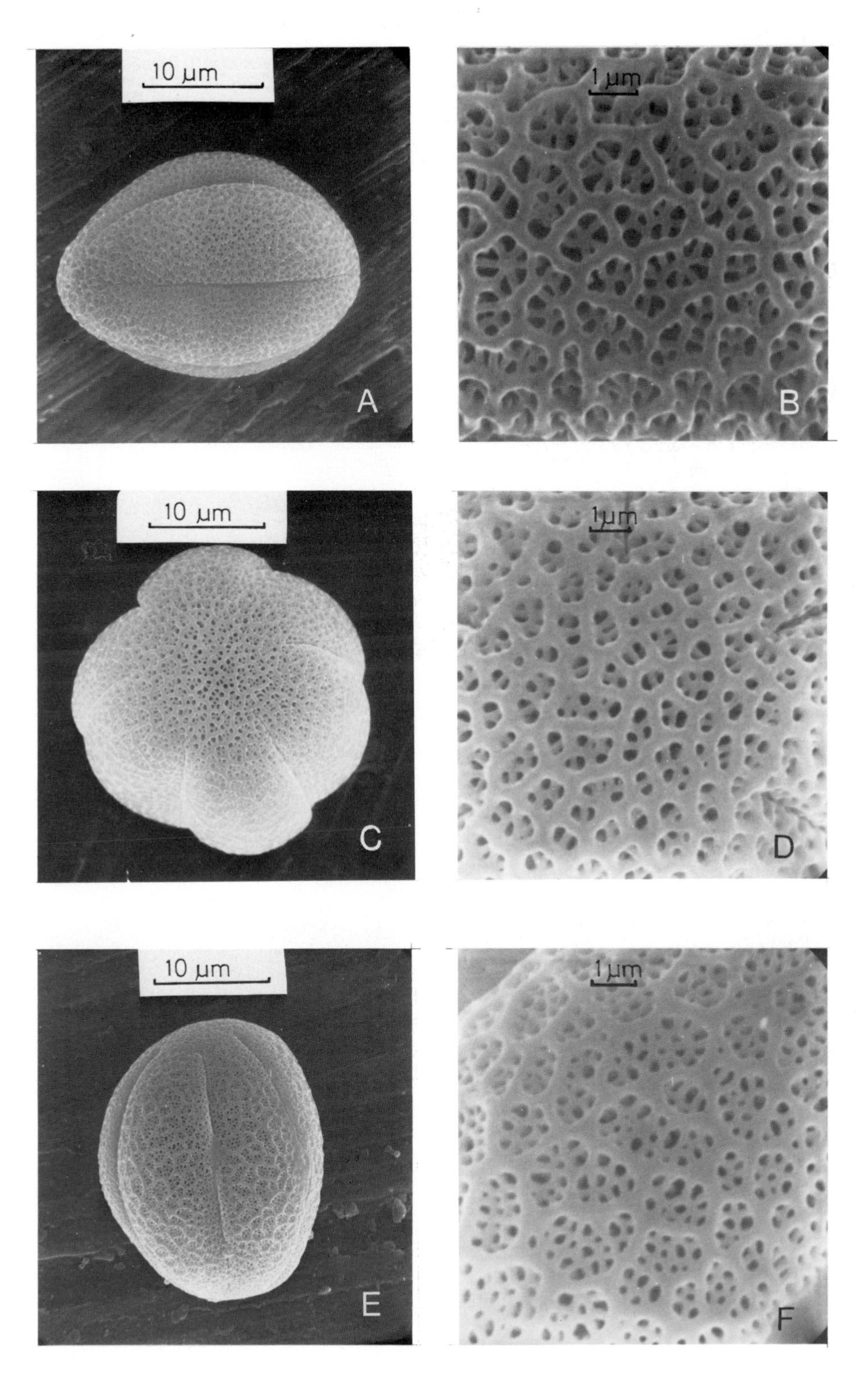

Plate 7

APPENDIX 1

List of Taxa Studied

Origanum L.

Section Origanum

O. compactum Benth.
O. ehrenbergii Boiss.
O. heracleoticum L.
O. isthmicum Danin
O. laevigatum Boiss.
O. ramonense Danin
O. virens Hoffmann & Link
O. viride (Boiss.) Hayek
O. vulgare L.

Section Majorana (Miller) T. Vogel

O. dubium Boiss.
O. majorana L.
O. micranthum T. Vogel
O. microphyllum Benth.
O. onites L.
O. syriacum L.

Section Amaracus Benth.

O. akhdarense Ietswaart & Baulus
O. amanum Post
O. cordifolium (Montbert & Aucher ex Bentham) Vogel
O. ciliatum Boiss. & Kotschy
O. dictamnus L.
O. hypericifolium Schwarz & Davis
O. leptocladum Boiss.
O. libanoticum Boiss.
O. lirium Held. ex Halácsy
O. rotundifolium Boiss.
O. saccatum Davis
O. scabrum Boiss. & Heldr.
O. sipyleum L.
O. tournefortii Aiton
O. vetteri Briq. & W. Barbey

Thymus L.

Section Thymus

T. capitatus (L.) Hoffmann & Link
T. ciliatus Benth.
T. vulgaris L.

Satureja L.

S. montana L.
S. obovata Lag.
S. thymbra L.

Micromeria Benth.

M. fruticosa (L.) Druce
M. nervosa (Desf.) Benth.

Hyssopus L.

H. officinalis L.
H. officinalis L. ssp. *montanus* (Jordan & Fourr.) Briq.

Melissa L.

M. officinalis L.

Lycopus L.

L. europaeus L.

Calamintha Miller

C. grandiflora (L.) Moench
C. sylvatica Bromf.
C. ascendens Jordan

CHAPTER 2

Ecology and Distribution

VOLATILE OILS AS ALLELOPATHIC AGENTS*

D. VOKOU, N.S. MARGARIS

1. INTRODUCTION

The term allelopathy was introduced by Molisch (1937) to cover all kinds of biochemical interactions among plants (microorganisms included) both of detrimental and beneficial effect. Later on the meaning of this term was restricted so as to refer only to harmful interferences.

The basic prerequisite of allelopathy is the escape of chemical compounds into the environment. Released by plants these chemical agents exert their deleterious effect in various ways.

Among the chemical substances most often implicated in allelopathic interferences are phenolics and terpenoids. The terpenoids build up the volatile oils of aromatic plants which are mostly found in the dry and hot areas occupied by phrygana, one type of the mediterranean climate ecosystems. The significant contribution of aromatic plants in such systems asks for elucidation of the ecological role of volatile oils. Various experiments provide indications upon which assumptions are formulated but no evidence exists at present to support a firm conclusion. The most important of these assumptions deals with functional relation of volatile oils with one or more of the following processes:

a. Antitranspirant action
b. Protection from phytophagous animals and insects
c. Interactions with microorganisms in the decomposition process
d. Contribution to the inflammability of mediterranean-type ecosystems
e. Selection of vegetative composition through allelopathic action

It is this last assumption we are dealing herewith.

Significant contribution to the recognition of allelopathy as an important element in vegetation patterning (10) had the outcome of the work carried out by Muller and his co-workers (9, 11, 13, 14) in the coastal sage

*This article is part of the Ph.D. Thesis of the first author.

Margaris N, Koedam A, and Vokou D (eds.): Aromatic Plants: Basic and Applied Aspects

ISBN 90-247-2720-0. Printed in the Netherlands.

communities of southern California, a region characterized by mediterranean type climate. The observation that bare zones, devoid of annuals, surrounded *Salvia* bushes initiated a series of experiments which proved the phytotoxicity not only of the volatile oil of *Salvia* species but of other aromatic plants, too, such as *Artemisia* ssp.(6, 7); the toxic potential of some volatile oil constituents was studied, too (12). Allelopathy induced by aromatic plants was also investigated in the climatically similar Israel (5).

In the phryganic ecosystems of Greece, the Greek homologue of the Californian coastal sage, aromatic plants contribute significantly both in terms of species number and of biomass. Since no relevant study was ever undertaken, we decided to test if the volatile oils of aromatic plants encountered in them exhibit any phytotoxic action and if they could exert through it any effect in ecological terms.

2. EXPERIMENTAL

The aromatic plants selected to provide the experimental material were *Thymus capitatus*, *Satureja thymbra*, *Teucrium polium* and *Rosmarinus officinalis*-the latter collected from a less xeric habitat. The material on which their effect was tested were *Cucumis sativus* and *Citrullus lanatus* seeds. *Cucumis sativus* has been successively used as experimental material in many related works. We decided to repeat the same test procedure with another species, *Citrullus lanatus*, for the sake of comparison.

Ten seeds of each species were let to germinate in a petri dish; in a small aluminum container inside the petri dish a quantity of the volatile oil under test was added in such a way that seeds and volatile oil were only in aerial contact. The petri dish was then closed with a plastic adhesive tape so as to avoid any escape to the environment and was put in the dark in $\simeq 20^{0}C$. For each volatile oil the quantities of 1, 3, 6, 10 and 20 μl were tested on both *Cucumis sativus* and *Citrullus lanatus* seeds and each treatment was replicated at least three times. Taking into account that the volume of the petri dish is approximately 100 cm^3 the quantities added correspond to concentrations of 10, 30, 60, 100 and 200 ppm, respectively.

As variables for the estimation of phytotoxic action both the percentage of germination and the total seedling length were chosen. In other related works the criterion of toxic action was the effect on the length of the seedling root. We have observed, however, that the whole seedling responds in the same way as the root alone-at least in our experiments. Therefore, we preferred it because it gives a better picture of the seedling potential for establishment and survival; further, in a more practical sense because it can express better the results of some of our experiments.

Having verified the toxicity of all volatile oils tested we proceeded in investigating at which stage of the germination process the toxic action is exerted.

First, we tested whether the toxic effect observed on the seedlings was due to seed damage at the dry state. For this reason seeds of *Cucumis sativus* were enclosed in petri dishes in an atmosphere of *Rosmarinus officinalis* volatile oil brought about with addition of the highest quantities used of 6, 10, and 20 µl. 50 seeds were used for each treatment. After 25 days the seeds were transferred into clean petri dishes, water was added to them and measurements were made after another 5 days.

Second, we tested whether the toxic effect is exerted at the pre- or post- germination stages (radicle emergence was taken as criterion of germination). Therefore, *Cucumis sativus* seeds were first let to germinate and were exposed afterwards to the influence of *Rosmarinus officinalis* volatile oil. According to their radicle length they were divided into five categories viz. of mean initial length of 0.1 cm, which coincides with simple radicle appearance, 0.3, 0.6, 0.9, and 1.8 cm and were treated separately as described above. The level of inhibition was estimated according to the following formula:

$$\%\ \text{inhibition} = \frac{\text{control seedling length} - (\text{final}-\text{initial})\text{seedling length}}{\text{control seedling length}} \times 100$$

In order to test whether the seedling damage induced is permanent or retrievable after removal of the toxic agent, inhibited *Cucumis sativus* seedlings, grown in the atmosphere of the most toxic quantities (6, 10, and 20 µl) of *Rosmarinus officinalis* volatile oil, with mean length of 0.1-0.8 cm, were transferred into clean petri dishes; their length was remeasured after seven days.

Experiments carried out as described above were repeated with seeds of wild annuals grown in the vicinity of aromatic plants in an attempt to approach natural conditions. The annuals were *Astragalus hamosus*, *Hymenocarpus circinatus* and *Medicago minima* and all four volatile oils were again tested on them. Experimentation with the seeds of these annuals became possible when germination arose from approximately 20% to ≃100% following scarification of the seed cover. 20 seeds were placed each time in petri dishes and each treatment was repeated at least three times.

Finally, the effect of the volatile oil of *Thymus capitatus* on its very seeds was investigated. In this case the lowest quantities of 1 µl, 3 µl, and 6 µl were used-the procedure being the same as above; additionally uncrushed leaves of *Thymus capitatus*, 0.5 g and 1 g, were also tested.

3. RESULTS AND DISCUSSION

In Table 1 there are data concerning the germination percentage of

Table 1. Germination percentage of *Cucumis sativus* and *Citrullus lanatus* seeds treated with 1, 3, 6, 10, and 20 µl of *Rosmarinus officinalis* (R), *Thymus capitatus* (T), *Satureja thymbra* (S), and *Teucrium polium* (TE), volatile oils (SD=control).

TREATMENT	GERMINATION (%)	
	Cucumis sativus	*Citrullus lanatus*
SD	85	94
R 1	90	87
R 3	83	93
R 6	93	93
R 10	95	97
R 20	83	83
T 1	94	91
T 3	91	93
T 6	87	80
T 10	87	87
T 20	90	83
S 1	89	87
S 3	87	91
S 6	87	94
S 10	94	94
S 20	87	91
TE 1	95	95
TE 3	87	83
TE 6	91	87
TE 10	91	91
TE 20	90	87

Cucumis sativus and *Citrullus lanatus* seeds under the different treatments imposed. It can be seen that it is in all cases very high which means that germination itself is not reduced. If, however, we consider the seedling growth, then an inhibitory effect is revealed. Fig. 1 presents data concerning *Cucumis sativus* seedling length under the different treatments imposed, expressed as per cent of inhibition of seedling growth. With the exception of the treatment: *Teucrium polium* volatile oil 1 µl, the difference from the control is significant at least at 0.05 level. It can be seen that the volatile oils of *Rosmarinus officinalis*, *Thymus capitatus* and *Satureja thymbra* when added in quantities of 6 µl, 10 µl, and 20 µl provoke a

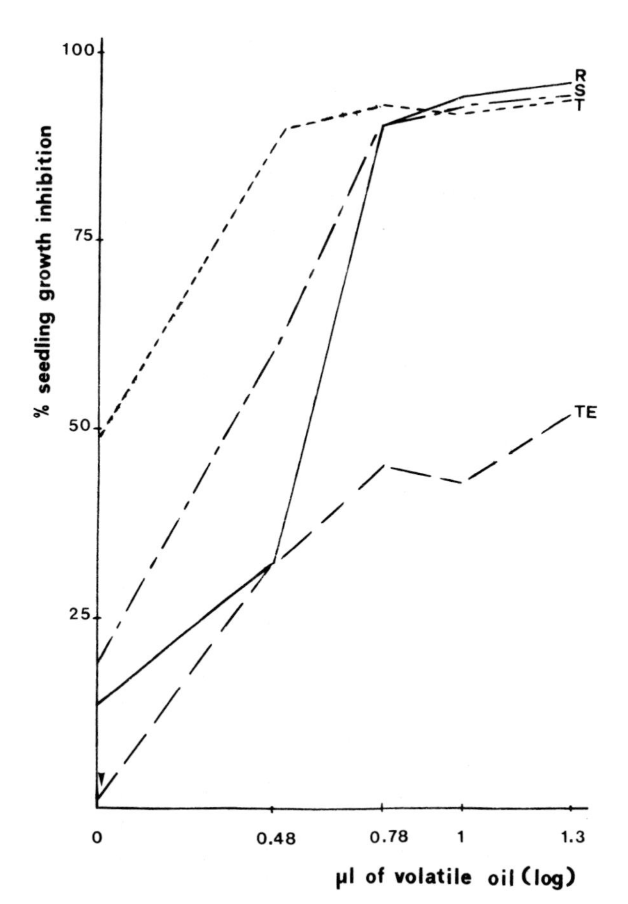

FIGURE 1. Growth inhibition of *Cucumis sativus* seedlings by 1, 3, 6, 10, and 20 µl of *Rosmarinus officinalis* (R), *Thymus capitatus* (T), *Satureja thymbra* (S), and *Teucrium polium* (TE) volatile oils (µl expressed in log; ▼ designates non-significant inhibition at p 0.05 level).

strong inhibitory effect of the seedling growth even rising to over 90%, while that of *Teucrium polium* is not as toxic as the others; nevertheless, it induces statistically significant inhibition of growth.

Figure 2 presents similar data concerning *Citrullus lanatus* seedlings. Only the means of the treatments: *Teucrium polium* 1 μl and *Satureja thymbra* 1 μl, do not differ significantly from the control at 0.05 level. As in the case of *Cucumis sativus* seedlings, here too, the volatile oils of *Rosmarinus officinalis*, *Thymus capitatus* and *Satureja thymbra* in quantities of 6 μl, 10 μl, and 20 μl strongly inhibit *Citrullus lanatus* seedlings (⩾ 80%) while the least toxic *Teucrium polium* volatile oil imposes a

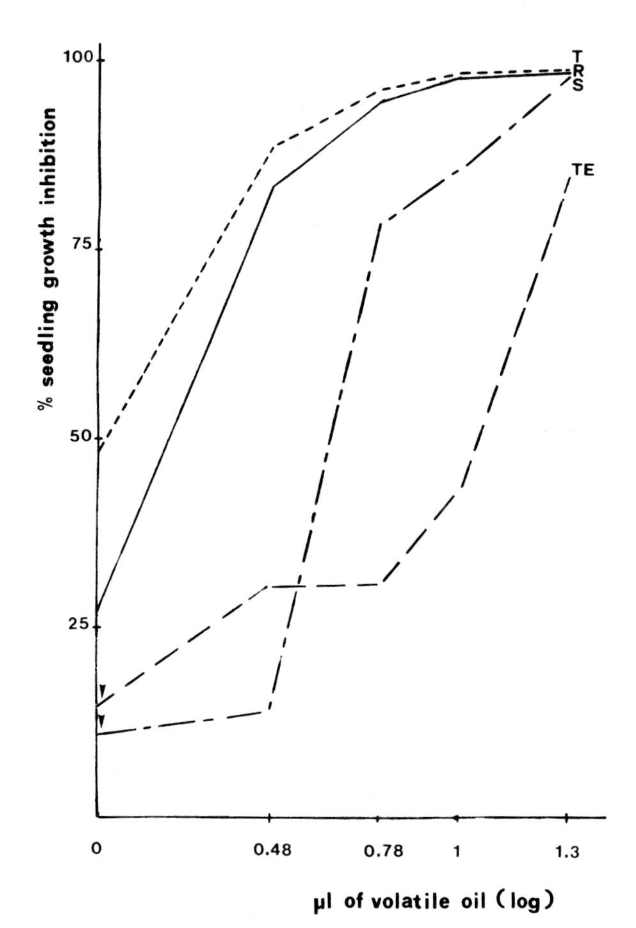

FIGURE 2. Growth inhibition of *Citrullus lanatus* seedlings induced by 1, 3, 6, 10, and 20 μl of *Rosmarinus officinalis* (R), *Thymus capitatus* (T), *Satureja thymbra* (S), and *Teucrium polium* (TE) volatile oils (μl expressed in log; ▼ designates non-significant inhibition at p 0.05 level).

TABLE 2. Effect of *Rosmarinus officinalis* volatile oil (6 μl, 10 μl, and 20 μl) on dry *Cucumis sativus* seeds. (➤ designates treatment imposed before watering and subsequent germination; SD_5=control of five days).

STAGES	TREATMENT	SEEDLING LENGTH (cm)
1. Dry *Cucumis sativus* seeds treated with R 6, R 10, R 20 for 25 days	SD_5	5.3±0.4
2. Transfer to clean atmosphere – soaking	R 6 ➤	5.3±0.4
3. Seedling length measurement (5 days)	R 10 ➤	6.3±0.5
	R 20 ➤	6.0±0.4

similar to the others inhibition only with the highest quantity of 20 μl.

As far as the stage where the toxic effect is imposed is concerned in Table 2 the results of the treatment where dry seeds of *Cucumis sativus* put into the atmosphere of 6 μl, 10 μl, and 20 μl of *Rosmarinus officinalis* volatile oil for 25 days and let to germinate afterwards show that all seeds thus treated germinated normally and the seedlings grew equally well as the untreated ones. This means that volatile oils are not likely to have any effect on the seeds at the dry state.

In Fig. 3, the bundle of curves -each one corresponding to a different initial seedling length, viz. of 0, 0.1, 0.3, 0.6, 0.9, and 1.8 cm- shows that there is no difference among the levels of inhibition imposed by each quantity of *Rosmarinus officinalis* volatile oil on the different categories of *Cucumis sativus* seedlings and seeds. These results seem to denote that whatever the effect of volatile oils, it is exerted at the post-germination stages -at least the first ones.

While in some cases the tissues of the inhibited seedlings, especially those grown in the atmosphere of 10 μl and 20 μl, seem to be seriously damaged the potential of recovery is retained and is evidently expressed when the toxic agent is removed. In Table 3 it can be seen that the inhibited *Cucumis sativus* seedlings of the treatments: 6 μl and 10 μl of *Rosmarinus officinalis* volatile oil, have totally recovered after seven days with their mean length not differing from the untreated seven-days-seedlings, while those of the treatment: 20 μl though differing significantly have started to grow with obvious signs of quick recovery. The first sign of recovery is the emergence of lateral roots taking the place

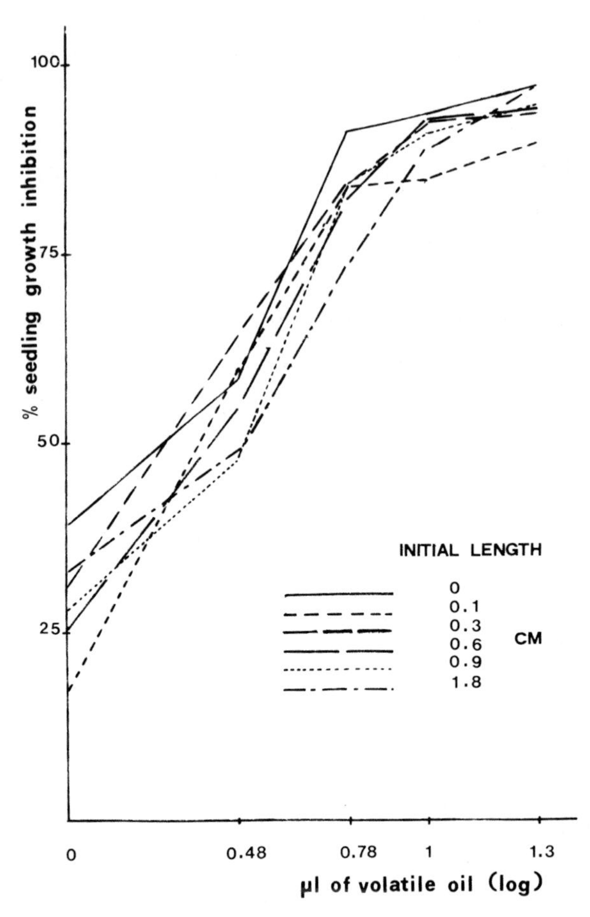

FIGURE 3. Comparative seedling growth inhibition of non- and pregerminated *Cucumis sativus* seeds treated with 1, 3, 6, 10, and 20 µl of *Rosmarinus officinalis* volatile oil. Pregerminated seeds were divided into five groups according to their initial seedling length, viz. 0.1, 0.3, 0.6, 0.9, and 1.8 cm; each group was further treated separately (µl expressed in log)

TABLE 3. Potential of recovery of inhibited *Cucumis sativus* seedlings pretreated with the most toxic quantities of 6, 10, and 20 µl of *Rosmarinus officinalis* volatile oil. (➤ designates treatment imposed before transfer far from the toxic agent; SD_7=control of seven days).

STAGES	TREATMENT	SEEDLING LENGTH (cm)
1. *Cucumis sativus* seeds treated with R 6, R 10, R 20. Seedling length after 5 days: 0.1 – 0.8 cm	SD_7	7.1 ± 0.5
	R 6 ➤	7.7 ± 0.5
2. Transfer to clean atmosphere	R 10 ➤	7.0 ± 0.4
3. Seedling length measurement (7 days)	R 20 ➤	3.3 ± 0.3

of the first and main radicle which is completely destroyed.

As far as the effect of volatile oils on wild annuals is concerned, results in Fig. 4 referring to *Hymenocarpus circinatus* show that the volatile oil of *Rosmarinus officinalis* proves to be the most toxic, provoking the strongest inhibition of seedling growth under all quantities used. *Satureja thymbra* and *Thymus capitatus* volatile oils which had a similar effect with that of *Rosmarinus officinalis* on the cultivated species provoked in this case a much reduced inhibitory effect attaining only 45-55% even at the highest added quantity of 20 µl. Surprisingly enough, the volatile oil of *Teucrium polium*, least toxic on cultivated species, induced a strong inhibition on this wild annual, greater than that of *Satureja*

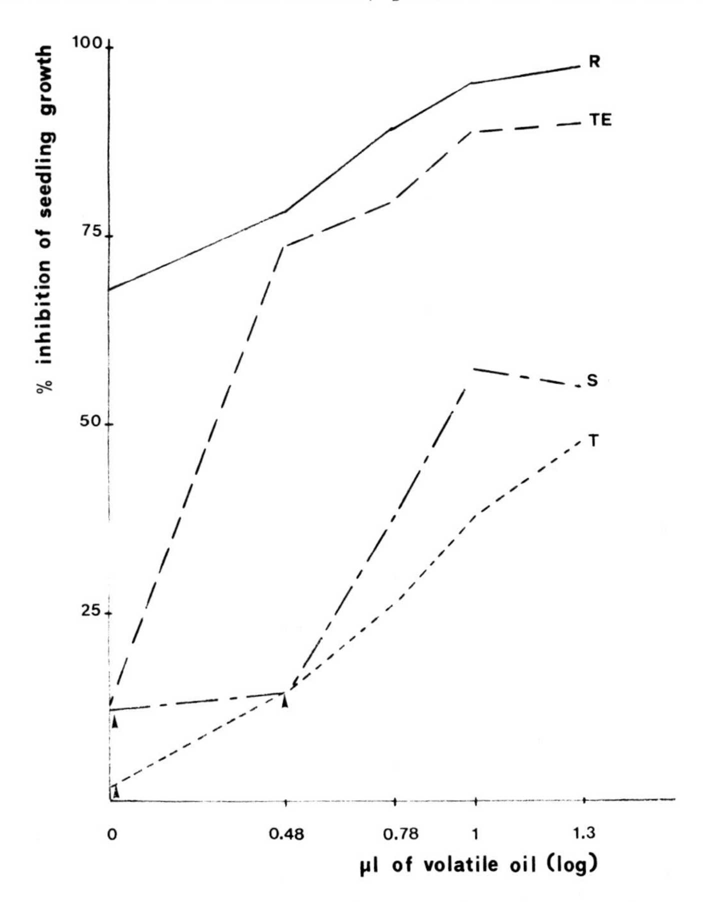

FIGURE 4. Growth inhibition of *Hymenocarpus circinatus* seedlings induced by 1, 3, 6, 10, and 20 µl of *Rosmarinus officinalis* (R), *Thymus capitatus* (T), *Satureja thymbra* (S) and *Teucrium polium* (TE) volatile oils (µl expressed in log; ▲ designates non-significant inhibition at p 0.05 level).

thymbra and *Thymus capitatus* volatile oils. Except for the treatments: *Thymus capitatus* 1 µl and 3 µl, and *Satureja thymbra* 1 µl all others differ significantly from the control at least at p 0.05 level.

Figure 5, referring to *Astragalus hamosus*, is similar to Fig. 4, concerning *Hymenocarpus circinatus*. Here, again, the volatile oil of *Rosmarinus officinalis* exerts the strongest inhibition while those of *Satureja thymbra* and *Thymus capitatus*, with a lesser effect, rise the level of inhibition to approximately 75% at the highest added quantity of 20 µl, and that of *Teucrium polium* proves, again, to be significantly toxic. With the exception of the treatment: *Teucrium polium* 1 µl, all others differ significantly from the control at p 0.05 level.

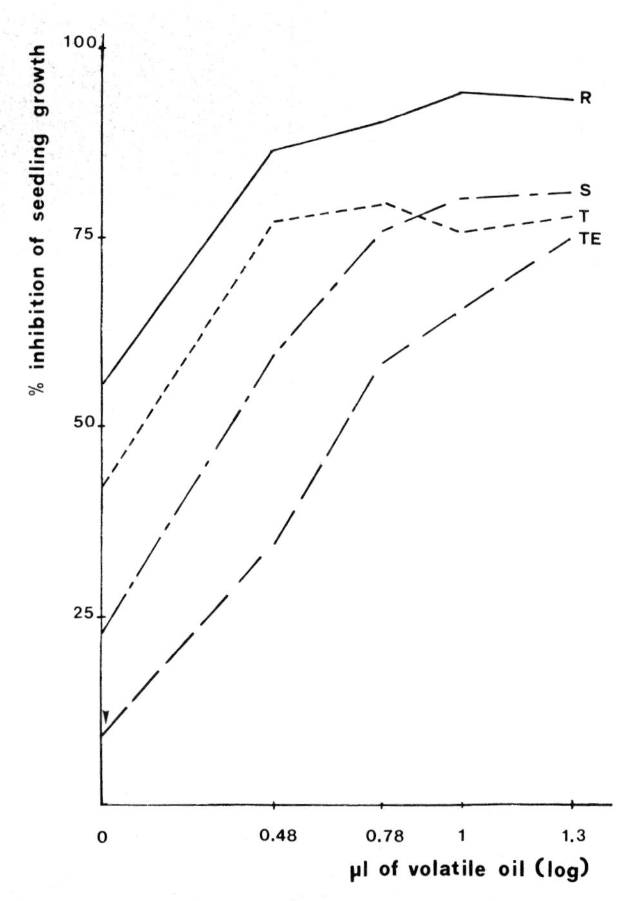

FIGURE 5. Growth inhibition of *Astragalus hamosus* seedlings induced by 1, 3, 6, 10, and 20 µl of *Rosmarinus officinalis* (R), *Thymus capitatus* (T), *Satureja thymbra* (S) and *Teucrium polium* (TE) volatile oils (µl expressed in log; ▼ designates non-significant inhibition at p 0.05 level).

The general picture of Fig. 6 concerning *Medicago minima*, is similar to Figs. 4 and 5 with *Rosmarinus officinalis* volatile oil being significantly more toxic than the others, followed by that of *Teucrium polium*, while those of *Satureja thymbra* and *Thymus capitatus* provoke an inhibition of 70-80% when the greatest quantity of 20 μl is added. With the exception of the treatment: *Satureja thymbra* 1 μl, the means of all others differ significantly at p 0.05 level.

As far as the effect on germination itself is concerned results inserted in Table 4 show that in principle it is not affected and seeds germinate in all cases quite normally.

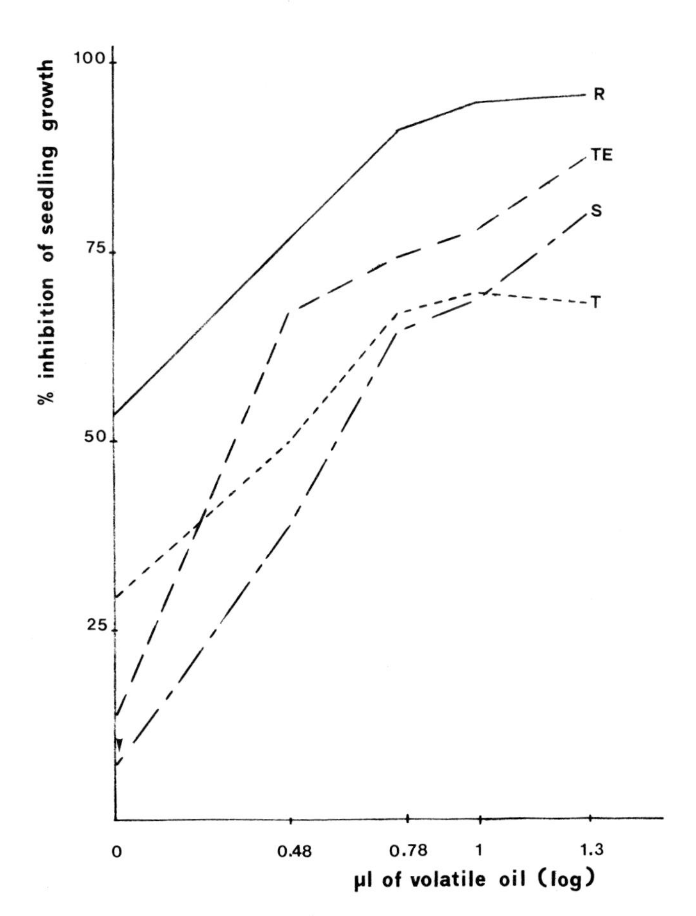

FIGURE 6. Growth inhibition of *Medicago minima* seedlings induced by 1, 3, 6, 10, and 20 μl of *Rosmarinus officinalis* (R), *Thymus capitatus* (T), *Satureja thymbra* (S), and *Teucrium polium* (TE) volatile oils (μl expressed in log; ▼ designates non-significant inhibition at p 0.05 level).

Table 4. Germination percentage of *Hymenocarpus circinatus*, *Astragalus hamosus* and *Medicago minima* seeds treated with 1, 3, 6, 10, and 20 µl of *Rosmarinus officinalis* (R), *Thymus capitatus* (T), *Satureja thymbra* (S), and *Teucrium polium* (TE) volatile oils.

TREATMENT	GERMINATION (%)		
	Hymenocarpus circinatus	*Astragalus hamosus*	*Medicago minima*
SD	93	92	98
R 1	90	93	92
R 3	97	76	93
R 6	95	67	95
R 10	93	62	76
R 20	48	56	63
T 1	92	96	98
T 3	88	84	92
T 6	85	72	90
T 10	92	72	95
T 20	93	48	94
S 1	89	85	99
S 3	95	82	89
S 6	87	69	86
S 10	71	83	83
S 20	61	78	69
TE 1	97	98	96
TE 3	88	93	90
TE 6	91	96	100
TE 10	94	96	100
TE 20	91	91	90

It must be mentioned that *Thymus capitatus* is the predominant aromatic plant in the phryganic ecosystem of Mt Hymettus, where the plants were collected from. If we compare the levels of inhibition imposed by volatile oils on cultivated and wild plants we see that the wild ones are somehow adapted to cope with the inhibitory action of the volatile oils of the dominant aromatic plants of the same system. On the contrary, such an adaptation is not evident in the case of the volatile oil of *Rosmarinus officinalis*, which is not a component of the aromatic flora of Hymettus phryganic ecosystem.

These results concerning *Rosmarinus officinalis* may give a different interpretation to the data provided by Deleuil (2, 3, 4) concerning absence of therophytes in the french garrigue with predominance of *Rosmarinus officinalis* and *Erica* species, which was attributed to toxic substances excreted by *Rosmarinus officinalis* roots. Volatile oils may have a contribution to this phenomenon.

Finally, the test of the effect of the volatile oil on the seeds of *Thymus capitatus* has revealed a remarkable example of autoallelopathy. In this case, not only seedling growth but germination itself was also affected. What is most important is that the minimum quantity of 1 μl, with little effect in all the other experiments conducted, provoked seedling growth inhibition of the level of approximately 83%. Similarly, germination of seeds and seedling growth of *Thymus capitatus* are much reduced in the presence of its leaves, what is due to the volatile oils emanated. Results are inserted in Table 5.

That this suppression of seedlings establishment really occurs in the field is supported by data provided by Argyris (1977). While *Thymus capitatus* seed production is very high, seedlings are very scarce in the phryganic ecosystem of Mt Hymettus, where he made his observations and measurements.

Restricted to establish and grow away from the mother plant the new individuals avoid competition which might be otherwise harmful for them. Furthermore, this interesting case of autoallelopathy could be interpreted

Table 5. Inhibition of *Thymus capitatus* seed germination and seedling growth imposed by *Thymus capitatus* volatile oil (TVO) - 1 μl, 3 μl, 6 μl - and leaves (TL) - 0.5 g and 1 g.

TREATMENT	GERMINATION (%)	INHIBITION OF SEEDLING GROWTH (%)
SD	73	
T VO 1	64	81
T VO 3	24	94
T VO 6	25	94
T L 0.5	44	88
T L 1	46	94

in terms of fire induced adaptations of plants. Fire is an inherent factor of the phryganic ecosystems, where this plant grows, frequently recurring. By destroying terpenic phytotoxins in the litter and soil, fire enables seeds of the soil seed bank to germinate and seedlings to establish thus securing survival of plants and recovery of the burnt system.

ACKNOWLEDGEMENT

The first author wishes to express her sincere thanks to A. ONASIS FOUNDATION for the financial support it provided to her.

REFERENCES

1. Argyris JP. 1977. Seed ecology of some phryganic species. Ph.D.Thesis, University of Athens (in Greek with english summary).
2. Deleuil MG. 1950. C. R. Acad. Sci. 230:1362.
3. Deleuil MG. 1951. C. R. Acad. Sci. 232:2038.
4. Deleuil MG. 1951a. C. R. Acad. Sci. 232:2476.
5. Friedman J, Orshan G, Ziger-Cfir Y. 1977. *J. Ecol.* 65:413.
6. Kelsey RG, Stevenson TT, Scholl JP, Watson TJ Jr, Shafizadeh F. 1978. *Biochem. Syst. Ecol.* 6:193.
7. McCahom CB, Kelsey RG, Sheridan RP, Shafizadeh F. 1973. *Bull. Torrey Bot. Club* 100:23.
8. Molisch H. 1937. Der Einfluss einer Pflanze auf die andere Allelopathie. Jena, Fischer (cited by Rice EL, 1974. Allelopathy. New York, San Francisco, London, Academic Press).
9. Muller CH. 1966. *Bull. Torrey Bot. Club* 93:332.
10. Muller CH, Muller WH, Haines BL. 1964. *Science* 143:471.
11. Muller WH. 1965. *Bot. Gaz.* 126:195.
12. Muller WH, Hauge R. 1967. *Bull. Torrey Bot. Club* 94:182.
13. Muller WH, Lorber P, Haley B, Johnson K. 1969. *Bull. Torrey Bot. Club* 96:89.
14. Muller WH, Muller CH. 1964. *Bull. Torrey Bot. Club* 91:327.

ARTEMISIA TRIDENTATA MONOTERPENOID EFFECT ON RUMINANT DIGESTION AND FORAGE SELECTION

B.L. WELCH, H. NARJISSE, E.D. McARTHUR

1. INTRODUCTION

Big sagebrush, *Artemisia tridentata* Nutt., is one of North America's most widespread, aromatic plant species. It dominates tens of millions of hectares and is present on millions more in western North America (30°-50° N latitude, 103°-123° W longitude) (3, 22).

A. tridentata is the dominant or preeminent species of the 11 species of the subgenus *Tridentatae* in terms of distribution, size (usually 0.5-2.0 m, but to 4.5 m tall), and total forage use (3, 22, 23, 24). *Tridentatae* are endemic to western North America. Until recently, *Tridentatae* were included as a section in the large North African-Eurasian subgenus *Seriphidium*. Geographical, karyotypic, growth habit, and chemotaxonomic considerations (15, 18, 22) led to the separation of *Tridentatae* from *Seriphidium* and elevation to independent subgeneric status (24). Despite the present importance of *A. tridentata* and its subgeneric allies in the western North American flora, they are relative newcomers. They have become important only since late Tertiary or early Quaternary periods and apparently diversified and expanded their ranges under the stimuli of alternating moist and dry climates during the Pleistocene Epoch (2, 22).

A. tridentata is comprised of 3 subspecies and 2 or 3 subordinate forms. The subspecies *tridentata, vaseyana,* and *wyomingensis,* are separated from one another by differences in habitat, leaf morphology, plant height, chemical constituents, rooting characteristics, and floral phenology (4,22). Subspecific distinctions are, however, sometimes difficult to make and, on some sites, introgression between subspecies is evident. Evidence from natural populations and controlled hybridization experiments among the subspecies and between them and other *Tridentatae* taxa indicates that hybridization occurs in the subgenus (23, 56, 58).

Tridentatae are an autopolyploid complex based on x=9. *A. tridentata* is mainly 2x and 4x. The subspecies *wyomingensis* is 4x with occasional 6x plants. The other 2 subspecies are comprised of both 2x and 4x populations

Margaris N, Koedam A, and Vokou D (eds.): Aromatic Plants: Basic and Applied Aspects

ISBN 90-247-2720-0.
Printed in the Netherlands.

with occasional mixed ploidy population (24). Tall populations are 2x (25). Populations of big sagebrush are often quite homogenous in morphological and chemical characteristics (4, 59, 60). Nevertheless, individuals within a population can be quite variable in growth characteristics (25) and chemical constituents (34). *A. tridentata* is ordinarily wind pollinated. Self pollination, however, appears to be possible at the cost of reduced seed yield (23, 58).

With its vast distributional range and large contribution to ground cover, big sagebrush is important in maintaining soil stability over a wide suite of sites. It is finding increasing use as a soil binder (10) on severely disturbed sites, such as mine spoils. The species is also used for ornamental purposes and has been used as a fuel. Its biggest use, however, is as a forage for wild ungulates, mule deer (*Odocoileus hemionus*) and pronghorn antelope (*Antilocapra americana*). It is also used to a lesser extent, by domestic livestock, but its forage value is often underestimated. Also, the presence of monoterpenoids may adversely affect an animal's digestion and forage choice.

2. DIGESTIBILITY

It has been known for some time that monoterpenoids have bacteriostatic and bactericidal activity (33, 35). In fact, microbiologists have actively investigated the bacteriostatic and bactericidal properties of monoterpenoids with the idea of determining their usefulness as prophylactic agents. With the discovery of antibiotics, the interest in using monoterpenoids to prevent or ward off bacterial infections died (33). Because of the antimicrobial nature of monoterpenoids, several researchers became concerned about the possible adverse effects monoterpenoid-producing plants might have on ruminant digestion (31, 36, 37, 44, 55).

Ruminant digestion depends heavily on the ability of rumen microorganisms to digest the coarse foods usually consumed by ruminants (21). Volatile fatty acids, the end product of rumen microbial fermentation, supply 60 to 80% of the energy used by ruminants (33). Thus, any compound that interferes with the rumen microbial activity could greatly reduce the supply of energy to the animal.

Nagy et al. (31), using *in vitro* techniques, found that the monoterpenoids of big sagebrush suppressed the growth of microorganisms in the mule deer rumen. Monoterpenoids also decreased the rate of cellulose digestion and the production of gas and volatile fatty acids by the rumen microorganisms. According to their report, cellulose digestion would be "slowed down slightly" when big sagebrush makes up 15-30% of the diet. They also suggested that rumen microorganisms might be able to adapt to the presence of big sagebrush monoterpenoids when animals continuously feed on big sagebrush. Oh et al. (36, 37) reported that monoterpenoid hydrocarbons (the monoterpenes of Douglas-fir, *Pseudotsuga menziesii*) actually enhanced *in vitro* microbial fermentation of sheep and deer rumen microorganisms. Sesquiterpenes were also found to be stimulatory. For the oxygenated monoterpenoids (alcohols, esters, aldehydes) microbial fermentation was inhibited. Oh et al. (36) reported that citronellal, an oxygenated monoterpenoid (aldehyde), inhibited rumen microorganisms from sheep and deer having no access to Douglas-fir (a monoterpenoid source), but produced no effect on rumen microorganisms from deer having access to Douglas-fir. This observation would support the Nagy group's idea that rumen microorganisms may be able to adapt to the presence of monoterpenoids (31); although, in a later report, Nagy and Tengerdy (32) noted no apparent adaptation of rumen microorganisms to the presence of big sagebrush monoterpenoids. (This interpretation of the Nagy and Tengerdy data (32) has been challenged by other scientists (58)).

Dietz and Nagy (13) suggest that the "theoretical"* decline of mule deer in the western United States is due principally to undue dependence on big sagebrush, juniper (*Juniperus* spp.), pine (*Pinus* spp.), and other species that contain high levels of monoterpenoids. They labeled monoterpenoids as toxic materials. Nagy (30) observed that when a rumen-fistulated goat was given increasing amounts of monoterpenoids, it went off feed when the concentration reached about 15 μl per 10 ml of rumen fluid. Nagy (30) reported that during a 30-day grazing trial deer experienced considerable weight loss and some became ill at the peak of big sagebrush consumption. Carpenter et al. (7) found, however, during a 30-day grazing trial with 6 mule deer, that at the peak of big sagebrush consumption, 3 deer were gaining weight and 2 deer were maintaining their weight. The sixth deer lost weight throughout the entire study. Wallmo et al. (55) embracing the *in vitro* evidence that big sagebrush monoterpenoids are toxic to rumen microorganisms describes an inescapable

*We term it a "theoretical" decline because some scientists did not believe that mule deer populations were declining or that the methods used to measure the decline were accurate (14, 54).

nutritional dilemma that faces wintering mule deer. The dilemma centers around the need of wintering mule deer to extract 7% crude protein from forages with an average crude protein content of 5%. "To increase protein intake, deer would need to consume more of the highly lignified browse twigs or more big sagebrush along with its toxic monoterpenoids, thereby lowering total digestibility". To increase digestibility, deer would need to eat more grass, which, in turn, would lower protein.

After considering the *in vitro* evidences that show big sagebrush monoterpenoids can suppress rumen microorganisms, we designed a study to test the feasibility of selection and/or breeding for strains of big sagebrush with low monoterpenoid content. We found that some accessions of big sagebrush contained higher levels of midwinter monoterpenoids than others (0.93%-2.95% dry matter), therefore, we felt that through selection and breeding schemes a low monoterpenoid strain of big sagebrush could be developed (58, 60). After reviewing the literature dealing with the digestibility of big sagebrush, we began to question whether monoterpenoids really suppress digestion in mule deer.

We found that neither *in vitro* nor *in vivo* digestibility trials using big sagebrush supported the contention that monoterpenoids interfere with the digestion of mule deer (6, 11, 19, 34, 40, 45, 46, 47, 48, 52). *In vitro* dry matter digestibility of big sagebrush was the second higher of all shrub species tested (Table 1). Caution must be used in drawing inference from the data presented in Table 1. Wallmo et al. (55) pointed out that the preparatory techniques used could have caused large losses of monoterpenoids from big sagebrush samples, thus introducing bias.

In vivo digestibility trials of big sagebrush showed that, for mule deer, it ranked second only to curlleaf mahogany in total digestible nutrients, even after correcting for monoterpenoids (Table 2). Probably smaller amounts of monoterpenoids were lost from the big sagebrush tissues fed to deer than were lost during *in vitro* trials. Smith (46) and Bissell et al. (6) provided freshly cut big sagebrush with unground stems and leaves to their deer on a daily basis. This would tend to minimize the loss of monoterpenoids. Dietz et al. (11) collected a big sagebrush sample in one large bundle. The sample was roughground in a combination hammermill and forage chopper and then placed in cold storage. Although this technique would result in a greater loss of monoterpenoids than the technique used by Smith (46) and Bissell et al. (6), it is less severe than the preparation of big sagebrush samples for *in vitro* digestibility

Table 1. Mean *in vitro* digestibility of shrub dry matter by mule deer inoculum. Data expressed as percent of digested dry matter.

Shrub	Percent of digested dry matter	Reference
	%	
Aspen	57.4	12
Big sagebrush	56.2	19, 34, 45, 52, 55, 63
Rose	54.5	12
Serviceberry	54.4	12
Curlleaf mahogany	53.5	52
Chokecherry	51.3	12
Russet buffaloberry	49.6	40
Willow	46.5	40
Snowberry	41.0	12
Blueberry	33.3	40
Bitterbrush	30.0	52
Mountain mahogany	28.5	52
Gambel oak	27.8	19

trials. None of these *in vivo* trials support the hypothesis that monoterpenoids interfere with digestion.

The conflict between *in vitro* or *in vivo* digestion trials and the *in vitro* studies that show that big sagebrush monoterpenoids can suppress rumen microorganisms led Welch and Pederson (63) to conduct a digestive study. This study was designed to overcome the errors of the *in vitro* digestion preparatory techniques used by other workers and to relate monoterpenoid content to digestibility. They were able to preserve the monoterpenoid content of big sagebrush by grinding the samples under liquid nitrogen with a steel motorized mortar and pestle (60, 63). Vegetative samples were collected in midwinter from 9 accessions of big sagebrush grown in a uniform garden. Other browse species were included for comparisons. Inoculum was collected from wild mule deer living on winter ranges dominated by big sagebrush. This was done to enhance the probability that the deer harvested had been eating big sagebrush long enough to have allowed the rumen microorganisms to adapt fully to the monoterpenoids (31, 32, 36). The digestion results of this study are

Table 2. *In vivo* digestibility of shrubs by mule deer.

Shrub	TDN*		Reference
Curlleaf mahogany	64.8		47, 48, 52
Big sagebrush	63.4	58.9**	6, 11, 46, 48, 52
Mountain mahogany	48.4		11, 48, 52
Cliffrose	47.2		48
Bitterbrush	46.0		6, 11, 47, 48, 52
Chokecherry	38.9		48
Oak	36.2		48

*TDN - Total digestible nutrients
**Corrected TDN value for monoterpenoids (58)

given in Table 3. The results indicate that big sagebrush is a highly digestible browse for wintering mule deer. Work cited earlier (6, 11, 19, 34, 45, 46, 48, 52) and a similar study conducted by Narjisse (34) with sheep agrees with these results.

In relating monoterpenoid content with digestibility, Welch and Pederson (63) found no relationship between digestibility and total monoterpenoid content (Table 4). Conolly et al. (9) obtained similar results using Douglas-fir and deer rumen inocula. On an individual monoterpenoid basis, camphene and camphor had a significant negative effect on digestion. Welch and Pederson (63) were unable to explain the negative effect that camphene and camphor appeared to have on digestibility among the 9 accessions of big sagebrush. Milford, Loa, and Clear Creek accessions contained about the same amount of camphor, but the Clear Creek accession was significantly more digestible than the Milford and Loa accessions. Trough Springs, Sardine Canyon, and Kaibab accessions contained the same amount of camphene, but the Kaibab accession was significantly more digestible than the Trough Springs accession. Their reluctance to accept the statistically significant effects of camphene and camphor on digestion stems from 2 points: 1) they selected the accessions purposely for varying amounts of total monoterpenoids not for individual monoterpenoids; so their tests for individual monoterpenoids may not be valid, and 2) *in vitro* digestibility values for big sagebrush are among the highest recorded for shrubs (58).

Table 3. *In vitro* dry matter digestibility among 9 accessions of big sagebrush and 4 other browse species.

Accessions of *Artemisia tridentata* or other browse species	Percent of digested dry matter
Clear Creek (A.t.)*	64.8[a]***
Dove Creek (A.t.)	64.6[a]
Loa (A.t.)	57.0[b]
Indian Peaks (A.t.)	55.8[b]
Benmore (A.t.)	55.2[b]
Kaibab (A.t.)	54.9[b]
Milford (A.t.)	54.6[b]
Rose hip**	49.1[bc]
Sardine (A.t.)	48.7[bc]
Curlleaf**	44.7[c]
Trough Springs (A.t.)	44.6[c]
Mahogany**	20.0[d]
Bitterbrush**	19.8[d]

* A.t.=An accession of *Artemisia tridentata* grown in a uniform garden

** Rose hip = *Rosa eglanteria*, curlleaf=*Cecrocarpus ledifolius*, mahogany=*Cecrocarpus montanus*, bitterbrush=*Purshia tridentata*.

*** Values sharing the same letter are not significantly different (P=0.05)

Table 4. The relationship between percent digestible dry matter (*in vitro*) and total monoterpenoids (essential oils), and individual monoterpenoids of 9 accessions of *Artemisia tridentata* (big sagebrush). Data expressed as percent dry matter.

	D.D.M.*	T.M.**	α-Pinene	Camphene	1.8-Cineol	α-Thujone	β-Thujone	Camphor	Terpineol
Trough Springs	44.6	1.41	0.09	0.09	0.08	0.00	0.00	0.89	0.02
Sardine	48.7	1.74	.02	.11	.06	.25	.58	.62	.05
Milford	54.6	.99	.05	.03	.01	.13	.04	.22	.00
Kaibab	54.9	.93	.11	.11	.02	.00	.00	.63	.01
Benmore	55.2	2.89	.00	.00	.09	.96	.49	.05	.32
Indian Peaks	55.8	1.72	.00	.04	.12	.03	.02	.31	.01
Loa	57.0	1.91	.01	.03	.09	.13	.55	.28	.05
Dove Creek	64.6	1.70	.00	.00	.02	.07	.71	.13	.00
Clear Creek	64.8	.95	.00	.00	.02	.05	.31	.24	.01
r^2		0.01	0.31	0.56	0.16	0.01	0.15	0.56	0.01

*D.D.M. = Digested dry matter. **T.M. = Total monoterpenoids

Another concern of Welch and Pederson (63) was the fate of monoterpenoids during the digestibility trials. Therefore, they modified the trials by adding three digestive tubes that contained only buffer. To these tubes, they added specific amounts of α-pinene and d-camphor (2.5 μg/μl). These tubes were incubated and treated like other tubes, except that they did not contain plant tissue or rumen inoculum, nor did they receive the acid-pepsin and sodium carbonate treatments. After the first incubation period, Welch and Pederson (63) extracted the solution with absolute ether and used gas chromatography to detect any changes in the monoterpenoid content. None of the α-pinene was recovered; all was lost from the flasks. Camphor loss was 17.3%. A white condensate had formed a ring around the neck of each flask. Each ring was located about 70 mm above the surface of the digestion solution; the condensate was identified as camphor. The force that drove these compounds out of the digestion solution was heat. Apparently, 38.5°C, which is close to the normal body temperature of mule deer, is sufficient to volatilize the monoterpenoids (28). Welch and Pederson (63) hypothesized that monoterpenoid levels could be greatly reduced in the deer rumen in three ways: 1) through mastication and rumination, 2) volatilization by body heat and eructation, and 3) possible absorption through the rumen wall and excretion through the kidneys (1).

This hypothesis was tested by Cluff et al. (8) with wild mule deer and by White et al. (64) with pygmy rabbits (*Brachylagus idahoensis*). They found monoterpenoid levels in the rumens of mule deer or the stomachs of pygmy rabbits were only 20 to 23% of expected levels. The expected levels were calculated by determining the amount of monoterpenoid-producing plants in the ingesta and the amount of monoterpenoid contained in the ingested plants. The level of monoterpenoids found in the rumen or stomach ingesta at the time of sampling does not appear to be high enough to interfere with microbial activity. Narjisse (34) also reported large losses of monoterpenoids from the rumena of rumen-fistulated goats and sheep. He infused up to 3 g of monoterpenoids into the rumen of each test animal, waited 4 hours, and then withdrew rumen fluid. The fluid was extracted for monoterpenoids, but none were present. This large loss of monoterpenoids may help explain the conflict between *in vitro* evidence that big sagebrush monoterpenoids inhibit rumen microorganisms and digestive trials which show that big sagebrush is a highly digestible winter forage.

Additional evidence that a high level of big sagebrush in the diet of mule deer does not adversely affect digestion stems from a report by Tueller

(51). He reported that the diet of mule deer wintering (December 1966, March 1966 , and December 1967) in the Fox Mountain area of Nevada, contained 69% big sagebrush, whereas the diet of mule deer wintering (December 1966, March 1966, and December 1967) in the White Rock area of Nevada contained only 28%. While the amount of big sagebrush in the diet of the deer from the 2 areas differed by a factor of 2.5, the amount of tail fat (an indicator of body condition)was almost the same in the 2 deer herds (32.4% for Fox Mountain deer and 29.1% for White Rock deer). Mule deer with a tail fat of 30% are considered to be in good physical condition. It should be repeated that 69% big sagebrush in the diet of Fox Mountain deer is well above the level considered safe by some workers (7, 31, 55).

3. SELECTIVITY

During digestibility trials, Smith (46) noted that deer showed definite aversion to individual big sagebrush plants. This preference of mule deer for certain accessions and individual plants of big sagebrush has been observed in the field by a number of researchers (16, 23, 38, 43, 45, 50, 61, 65, 66). A study conducted by Welch et al. (61) demonstrated the differential preference of wintering mule deer for accessions of big sagebrush grown on a uniform garden (Table 5). Along this same line, Nagy and Tengerdy (32) and Dietz and Nagy (13) hypothesized that black sagebrush (*Artemisia nova*) is preferred over big sagebrush by mule deer. It was their contention that the black sagebrush contains lower monoterpenoid concentrations than big sagebrush, which would be less inhibitory to microorganisms. It was thought lower concentrations cause the deer to select black sagebrush over big sagebrush. Observations made by Smith (46) and Sheehy (45), however, showed that mule deer did not prefer black sagebrush over big sagebrush, even though Sheehy (45) found the monoterpenoid content to be lower in black sagebrush than big sagebrush.

Other attempts have been made to relate monoterpenoid content to preference. Sheehy (45) reported that the relative concentration of 8 monoterpenoids could account for 90% of the variation in mule deer utilization among 7 sagebrush taxa. He found it took 12 monoterpenoids to account for 90% of the variation in sheep utilization among 7 sagebrush taxa. He noted that monoterpenoids had only a minor negative influence in determining preference. Scholl et al. (43), however, using the relative concentration of 8 monoterpenoids, could only account for 20.7% of the variation in wild

Table 5. The winter preference of wild mule deer for 10 accessions of big sagebrush (*Artemisia tridentata*) grown on a uniform garden near Helper, Utah.

Winter of 1977		Winter of 1978	
Accessions	% Used	Accessions	% Used
Hobble Creek	84[a*]	Hobble Creek	74[a]
Spanish Valley	73[a]	Monticello	59[ab]
Monticello	63[ab]	Spanish Valley	54[ab]
Indian Peaks	59[b]	Dove Creek	50[b]
Milford	58[b]	Indian Peaks	47[b]
Indianola	47[bc]	Milford	45[bc]
Dove Creek	46[bcd]	Indianola	40[bc]
Loa	33[cd]	Loa	24[cd]
Trough Springs	29[d]	Marysvale	12[d]
Marysvale	19[d]	Trough Springs	10[d]

*Values sharing the same letter superscript are not significantly different at the 5% level.

mule deer utilization among 12 sagebrush taxa. Radwan and Crouch (39) found that the "families of Douglas-fir" varied significantly in yield and composition of monoterpenoid, but differences were not related to deer preference.

The strongest evidence that monoterpenoids influence animal preference comes from studies conducted by Schwartz et al. (44), and Narjisse (34). Schwartz et al. (44) used tame, nonexperienced mule deer as test animals in cafeteria feeding trials. The trials were designed to determine the preference of the test animals for 3 species of juniper and for pelleted feed treated with different levels of monoterpenoids. Their results showed that tame mule deer preferred food that had the lowest levels of oxygenated monoterpenes. In a test to determine the influence of the odour of monoterpenoids on food preference in sheep and goats, Narjisse (34) found that, for the first 2 days of the trial, sheep ate more food from the feed bins lacking the odour of monoterpenoids. Selection after the first 2 days was at random. Goats did not discriminate against the odour of monoterpenoids. Narjisse (34) using anosmic sheep and goats found that goats did discriminate against the taste of monoterpenoids. Sheep, however, showed no discrimination.

Welch et al. (62) related the preference of wild mule deer for 20 accessions of *Artemisia* spp. to total monoterpenoids and individual fractions of monoterpenoids. Their results indicated no significant relationship between preference and monoterpenoid content. Thus, 3 studies (34, 44, 45) support the contention that monoterpenoid levels influence test animal preference and 3 studies (39, 43, 62) do not. It may be that in the case of the cafeteria trials by Schwartz et al. (44) and Narjisse (34), where all other factors were held constant, that monoterpenoids significantly influenced preference and that under the field conditions, such as those in the Welch et al. (62) study, monoterpenoid influences may be masked by other factors.

One of these other factors could be animal experience (34, 67). Zimmerman (67) reported that in the case of raising cattle on shrub ranges of Nevada, USA, it is very important that calves stay with their mothers to learn how to survive on shrub ranges. Without this experience, it would be doubtful that the calves could survive on the shrub diet. Narjisse (34) in range tests reported that experienced range sheep consumed significantly higher levels of big sagebrush than inexperienced sheep. Carpenter et al. (7) found that tame mule deer also increased big sagebrush consumption over time. Unknown animal factors may play as important a role in selectivity as plant factors do.

Another plant factor that could play an important part in selectivity is the presence of secondary compounds other than monoterpenoids. Monoterpenoids are considered secondary compounds because they have no known function in the maintenance of plant life (29). The significance of secondary compounds has been considered by Bell (5) to be twofold: phylogenetic and ecological. As we have seen above, the ecological significance of monoterpenoids in animal selectivity remains an elusive phenomenon. Their value and that of sesquiterpene lactones (another important compound in unraveling phylogenetic relationships in *Artemisia* at the subgeneric level and below) (15, 18) may lie in chemotaxonomy. Sesquiterpene lactones from *Artemisia* have been shown to have a wide range of biological activities in man and other animals (41).

Other secondary compounds quite common in *Artemisia tridentata* include the phenolic compound classes flavonoids and coumarins. Both flavonoids and coumarins have been useful in solving taxonomic problems with *A. tridentata* and the *Tridentatae* (16, 17, 24). Coumarin compounds (principally isoscopoletins, scopoletin, and esculetin) have been associated with mule deer preference for some sagebrush taxa (50). We emphasize, however, that no direct link has been established between coumarins and preference. The coumarins may serve as hybridity and palatability markers in an improvement program for big sagebrush (58).

4. THE VALUE OF BIG SAGEBRUSH AS A FORAGE

The aromatic plant, big sagebrush, has several characteristics which make it a valuable browse for wintering mule deer and sheep. Its midwinter crude protein content of 10 to 12% of dry matter ranks among the highest for all range plants (51, 57, 59). Big sagebrush winter digestibility as measured by total digestible nutrients (56.8%) ranks just below curlleaf mahogany (*Cercocarpus ledifolius*) (64.8%), sand dropseed grass (*Sporobolus cryptandrus*) (59.0%), and western wheatgrass (*Agropyron smithii*) (57.6%) (57). Phosphorus content (0.20% of dry matter) of big sagebrush is the highest among range plants commonly found on winter ranges in the western United States (57). Carotene content of big sagebrush is about 4.6 times higher than the minimum requirement of most wintering ruminants (57).

Some woody *Artemisias* of Asia are also considered to be excellent winter feed. Larin (20) chronicled the value of *Artemisia* in general in the steppe area of the Soviet Union. Sal'manov and Khamdanov (42) pointed out that *A. halophilia* is a "complete nutritional feed" in the Uzbek SSR. McArthur and Harrington (26) documented the heavy use, high protein content, and high dry matter digestibility of *A. herba-alba* in Afghanistan.

Another important, but little recognized, characteristic of big sagebrush is its ability to maintain stable yields of forage during the years of abnormally low rainfall. This makes big sagebrush a more reliable winter forage source than other range plants (25, 27). One possible negative characteristic of big sagebrush was pointed out by Smith (49), namely, that penned deer when fed pure diets of big sagebrush restricted the amount of big sagebrush eaten, although Smith (46) was able to maintain deer for 2 months on a pure diet of big sagebrush without extreme weight loss. Feed restriction and weight loss during the winter may be a normal reaction associated with normal physiological changes (53). Wintering animals are probably better off having a varied diet. Because of its high protein, digestibility, phosphorus, and carotene content and its reliability as a forage source during drought, big sagebrush is a valuable winter browse for mule deer and sheep.

REFERENCES

1. Annison EF. 1965. In: Physiology of Digestion in the Ruminant, Dougherty RW (ed.), p. 185. Washington, Butterworths.
2. Axelrod DI. 1950. *Carnegie Inst. Wash. Publ.* 590:215.
3. Beetle AA. 1960. *Bull. Univ. Wyom. Exp. Stn.* 368:1
4. Beetle AA. 1971. *Madroño* 20:385.
5. Bell EA. 1980. In: Encyclopedia of Plant Physiology, New Series, Vol. 8. Secondary Plant Products, Bell EA and Charlwood BV (eds.). Berlin, Springer Verlag.
6. Bissell HD, Harris B, Strong H, James F. 1955. *Calif. Fish and Game* 41: 57.
7. Carpenter LH, Wallmo OC, Gill RB. 1979. *J. Range Manage.* 32:226.
8. Cluff LK, Welch BL, Pederson JC, Brotherson JD. 1982, in press. *J. Range Manage.* 35.
9. Connolly GE, Ellison BO, Fleming JW, Geng S, Kepner RE, Longhurst WM, Oh J, Russell GF. 1980. *For. Sci.* 26:179.
10. DePuit EJ, Stelter V (eds.). 1981. Proc. Shrub Establishment on Disturbed and Semi-arid Lands. Laramie, Univ. Wyoming Press.
11. Dietz DR, Udall RH, Yeager LE. 1962. *Colo. Fish and Game Dept. Tech. Publ.* 14:46.
12. Dietz DR. 1972. In: Wildland shrubs-their biology and utilization, Mckell CM, Blaisdell JP, Goodin JR (eds.), p. 289. USDA For. Serv. Gen. Tech. Rep. INT-1. Intermt. For. and Range Exp. Stn., Ogden, Utah.
13. Dietz DR, Nagy JG. 1976. In: Mule Deer Decline in the West: A Symp., Workman W, Low JB (eds.), p. 71. Coll. Nat. Resour., Utah Agric. Exp. Stn., Logan.
14. Gill RB.1976. In: Mule deer decline in the West: A Symp., Workman W, Low JB (eds.), p. 99. Coll. Nat. Resour., Utah Agric. Exp. Stn., Logan.
15. Greger H. 1978. In: The biology and chemistry of the Compositae, Heywood VH, Harborne JB, Turner L (eds.), p. 899. London, Academic Press.
16. Hanks DL, McArthur ED, Stevens R, Plummer AP. 1973. USDA For. Serv. Res. Pap. INT-141:1. Intermt. For. and Range Exp. Stn., Ogden, Utah.
17. Holbo HR, Mozingo HN. 1965. *Amer. J. Bot.* 52:970.
18. Kelsey RG, Shafizadeh F. 1979. *Phytochemistry* 18:1591.
19. Kufeld RC, Stevens M, Bowden DC. 1981. *J. Range Manage.* 34:149.
20. Larin IV. 1956. Pasture economy and meadow cultivation. Nat. Sci. Found. Publ., Washington, D.C. by the Israel Program for Scientific Translations.
21. Maynard LA, Lossli JK, Hintz HF, Warner RG. 1979. Animal nutrition, 7th ed. New York, McGraw-Hill Co.
22. McArthur ED, Plummer AP. 1978. *Great Basin Nat. Mem.* 2:229.
23. McArthur ED, Blauer AC, Plummer AP, Stevens R. 1979. USDA For. Serv. Res. Pap. INT-220, 82 p. Intermt. For. and Range Exp. Stn., Ogden, Utah.
24. McArthur ED, Pope CL, Freeman DC. 1981. *Amer. J. Bot.* 68:589.
25. McArthur ED, Welch BL. 1982, in press. *J. Range Manage.* 35.
26. McArthur ED, Harrington GN. 1978. In: Proc. First Int. Rangeland Cong., Hyder DN (ed.), p. 596. Soc. for Range Manage., Denver, Colo.
27. Medin DE, Anderson AE. 1979. *Wildl. Monographs* 68:1.
28. Moen AN. 1973. Wildlife ecology. San Francisco, Calif., W.H. Freeman Co.
29. Mothes K. 1980. In: Encyclopedia of Plant Physiology. New Series, Vol. 8, Secondary Plant Products, Bell EA, Charlwood BV (eds.), p. 1. Berlin, Springer-Verlag.
30. Nagy JG. 1979. In: Sagebrush ecosystem: A Symp., p. 164. Coll. Nat. Resour., Utah Agric. Exp. Stn., Logan.

31. Nagy JG, Steinhoff HW, Ward GW. 1964. *J. Wildl. Manage.* 28:785.
32. Nagy JG, Tengerdy RP. 1968. *Appl. Microbiol.* 16:441.
33. Nagy JG, Regelin WL. 1977. *Game Biol.* 13:225.
34. Narjisse H. 1981. Acceptability of big sagebrush to sheep and goats: Role of monoterpenes. Ph.D. Thesis. Logan, Utah State Univ.
35. Nicholas HJ. 1978. In: Phytochemistry, Vol. II. Organic metabolites, Miller LP (ed.). New York, Van Nostrand Reinhold Co.
36. Oh HK, Sakai T, Jones MB, Longhurst WM. 1967. *Appl. Microbiol.* 15:777.
37. Oh HK, Jones MB, Longhurst WM. 1968. *Appl. Microbiol.* 16:39.
38. Plummer AP, Christensen DR, Monsen SB. 1968. Restoring big game range in Utah. *Utah Fish and Game Publ.* 68-3.
39. Radwan MA, Crouch GL. 1978. *J. Chem. Ecol.* 4:675.
40. Regelin WL, Wallmo OC, Nagy JG, Dietz DR. 1974. *J. For.* 72:282.
41. Rodrigues E, Towers GHN, Mitchell JC. 1976. *Phytochemistry* 15:1573.
42. Sal'manov NS, Khamdamov IKh. 1978. *Rastit. Resur.* 14:3 (in Russian) (abstracted in *Biol. Abstr.* 67:5, ref. 45 (1979)).
43. Scholl JP, Kelsey RG, Shafizadeh F. 1977. *Biochem. System. Ecol.* 5:291.
44. Schwartz CC, Regelin WL, Nagy JG. 1980. *J. Wildl. Manage.* 44:114.
45. Sheehy DP. 1975. Relative palatability of seven *Artemisia* taxa to mule deer and sheep. M.S.Thesis, Corvallis, Oregon State Univ.
46. Smith AD. 1950. *J. Wildl. Manage.* 14:285.
47. Smith AD. 1952. *J. Wildl. Manage.* 16:309.
48. Smith AD. 1957. *J. Range Manage.* 10:162.
49. Smith AD. 1959. *J. Range Manage.* 12:8.
50. Stevens R, McArthur ED. 1974. *J. Range Manage.* 27:325.
51. Tueller PT. 1979. Food habits and nutrition of mule deer on Nevada ranges. Nevada Agric. Exp. Stn., Reno, Univ. Nev.
52. Urness PJ, Smith AD, Watkins RK. 1977. *J. Range Manage.* 30:119.
53. Urness PJ. 1980. Supplemental feeding of big game in Utah. Utah Div. Wildl. Resour. Publ. 80-8.
54. Vogt B. 1978. The big ones are back! Nat. Wildl. Conf. Oct-Nov. 5-12.
55. Wallmo OC, Carpenter LH, Regelin WL, Gill RB, Baker DL. 1977. *J. Range Manage.* 30:122.
56. Ward GH. 1953. *Contr. Dudley Herb.* 4:155.
57. Welch BL. 1981. In: Shrub establishment: A Symp., Stelter L, DePuit EJ, Mikol S (eds.), p. 9. Range Manage. Div., Laramie, Univ. Wyoming.
58. Welch BL, McArthur ED. 1979. In: Arid Land Plant Resources, Goodin JR, Northington K (eds.), p. 451. Inter. Center for Arid and Semiarid Land Studies. Lubbock, Texas Tech. Univ.
59. Welch BL, McArthur ED. 1979. *J. Range Manage.* 32:467.
60. Welch BL, McArthur ED. 1981. *J. Range Manage.* 34:380.
61. Welch BL, McArthur ED, Davis JN. 1981. *J. Range Manage.* 34:409.
62. Welch BL, McArthur ED, Davis JN. 1981. Unpublished data on file at the Shrub Sciences Lab., Provo, Utah.
63. Welch BL, Pederson JC. 1982, in press. *J. Range Manage.* 35.
64. White SM, Welch BL, Flinders JT. 1982, in press. *J. Range Manage.* 35.
65. Willms WA, McLean A, Tucker R, Ritcey R. 1979. *J. Range Manage.* 32:299.
66. Winward AH. 1970. Taxonomic and ecological relationships of the big sagebrush complex in Idaho. Ph.D. Thesis. Moscow, Univ. Idaho.
67. Zimmerman EA. 1980. *Rangelands* 2:184.

PRODUCTIVITY OF AROMATIC PLANTS: CLIMATIC MODELS

E.O. BOX

1. INTRODUCTION

Aromatic plants are part of a more general class of plants which emit a variety of secondary substances. The role of their volatile oils, composed of isoprene compounds, is not clear; they may contribute to regulating water loss, making plants more flammable or act as defense mechanisms. Aromatic plants, include a variety of primarily small shrubs (e.g. *Thymus, Cistus*) and some herbs (e.g. *Lamium, Mentha*) which are especially characteristic of dry mediterranean climates (i.e. garrigue, phrygana, or batha, as opposed to maquis). Some larger shrubs (e.g. *Myrtus, Rosmarinus*) and even small trees (e.g. *Juniperus*) may also be included. Such vegetation is usually considered to be a fire climax or subclimax, since it burns regularly. The frequency and intensity of such fires depend mainly on the amount of phytomass accumulated since the last fire and on its flammability.

Aromatic plants have had a wide variety of economic uses for many centuries, in areas such as cooking, medicine, cosmetics, fuel. Because of the inevitability of fire, with concomitant loss of potentially usable phytomass, it has been suggested that mediterranean-type vegetation be harvested, for energy and organics, on a regular basis. This would also reduce risks associated with naturally occurring fires (10). In order to know if harvesting is feasible, fire frequency, annual productivity, and recovery rates must be determined, as a basis for estimating how much and how often the vegetation can be harvested. Because of their dependency on climatic conditions, these rates can often be determined for particular areas and then predicted for other areas by means of climate-based models.

The purpose of this paper is to state some of the questions involved in harvesting or otherwise utilizing aromatic vegetation and then to examine some of the modeling possibilities which might be employed to provide the necessary information for rational vegetation management. Some basic questions can be stated easily:

- Is the regular harvesting of mediterranean vegetation feasible and

Margaris N, Koedam A, and Vokou D (eds.): Aromatic Plants: Basic and Applied Aspects

ISBN 90-247-2720-0. Printed in the Netherlands.

desirable (economically as well as ecologically)?
- Is the regular harvesting of particular aromatic species being carried out wisely (economically and ecologically)?
- Which types of vegetation/plants can be harvested?
- How much can be harvested at one time?
- How often can it be harvested (what are the recovery times)?

Information on aromatic plants may be obtainable not only from studies directed at mediterranean vegetation and harvesting but also from work on allelopathy and on secondary emissions by other types of plants, such as pines which have been implicated in connection with "air pollution" in some areas (6).

2. CLIMATIC RELATIONS OF EMITTING PLANTS

Plants which emit aromatic or similar secondary substances appear to be most characteristic of climates with significant but not extreme water stress during at least one season. This includes not only aromatic plants of mediterranean climates but also temperate/subtropical pines (hot summers with water stress on sunny days), various shrubs of dry regions (e.g. *Larrea tridentata*, the "creosote bush" of the Americas), and various trees and arborescents of the humid tropics (e.g. rubber trees, which undergo water stress for several hours around midday). Many other taxa emit less noticeable secondary substances in less summer-dry climates, including most conifers (mainly terpenes), oaks (isoprene), and even quite malacophyllous deciduous trees (14).

The structure of aromatic vegetation is suggested by the vegetation profiles shown in Table 1 for three sites in the Mediterranean region plus one in mediterranean Australia and one in the humid warm-temperate region of southeastern USA. These profiles are based on a world eco-physiognomic life-form classification and are generated by means of a generating program ECOSIEVE plus climatic envelopes for each of the 90 basic plant types (4). The estimated climatic envelopes of the most important life forms involving aromatic plants, plus a listing of other possible emitting forms, are shown in Table 2. The two most important plant types of mediterranean vegetation are the evergreen sclerophyll shrubs (and small trees) and the seasonally dimorphic dwarf-shrubs. With smaller leaf area the dwarf-shrubs tolerate drier climates but are limited to a shorter spring (and sometimes also autumn) growing season due to shallower root systems and greater water loss rates. These are the "mediterranean dwarf-shrubs" in Tables 1 and 2.

The most characteristic aromatic vegetation type is the phrygana (coastal sage in California) dominated by aromatic chamaephytes. Less dry mediterranean climates generally support denser shrublands (maquis, chaparral), while still moister mediterranean climates (general coastal or montane) may support sclerophyll forests (e.g. *Cedrus, Pinus, Cupressus, Eucalyptus*). Not all plants in any of these vegetation types will be aromatic. The flammability of such vegetation will depend on the vegetation composition as well as the degree of summer dryness.

3. PRODUCTIVITY MODELLING IN GENERAL

The modelling of plant and vegetation functional processes, including productivity, can be approached in two basic ways:

1. deterministic models, in which the rates or amounts are estimated directly (from environmental determinants) by simple, deterministic equations.
2. stochastic models, in which rates and amounts depend also on the current state of the vegetation.

In the first case the equations are generally derived by least-squares curve-fitting using a data-base which covers the full (geographic) ranges of variation of the environmental determinants and the functional values. The most successful environmental predictors for plant productivity and related processes appear to be climatic variables which express available energy and water, above all actual evapotranspiration (7, 9, 11). Such an approach has lead to various relatively simple, world-scale models for annual vegetation productivity (7, 8, 12). The simplest of these models, at least conceptually, is probably the so-called "Montréal Model" (7), in which annual net primary productivity (NPP, dry matter) is conceived to be a saturation function of actual evapotranspiration:

$$NPP = 3000 \left[1 - e^{-0.0009695\,(AET-20)} \right] \qquad (1)$$

Such models can easily be converted into predictive maps and even quantified by computer planimetry to obtain estimates of total amounts (1, 2, 3, and unpublished data).

The stochastic approach involves essentially all components of the deterministic but builds them into a system-simulation model which permits simulation over time (e.g. over the course of vegetation succession) and

```
    Location
 TMAX   TMIN   PRCP   PMAX   PMIN   PMTMAX   MI

IRAKLION/KRITI                 HELLAS                          27m
 26.0  11.8  533.   98.    1.    1.    0.59
       * 1. MEDITERRANEAN EVERGREEN SHRUBS                 MI      0.25
       + 2. SHORT BUNCH-GRASSES                            TMAX    0.50
         3. NEEDLE-LEAVED EVERGREEN SHRUBS                 MI      0.33
         4. HOT-DESERT EVERGREEN SHRUBS                    TMAX    0.22
         5. XERIC EVERGREEN TUFT-TREELETS                  TMIN    0.17
       + 6. MEDITERRANEAN DWARF-SHRUBS                     TMAX    0.47
         7. Various STEM-SUCCULENTS
       + 8. XERIC CUSHION-SHRUBS                           TMAX    0.32
         9. Various FORBS

TARIFA                         ESPAÑA                          46m
 23.0  12.0  683.  120.    0.    5.    0.86
         1. MEDITERRANEAN BROAD-EVERGREEN TREES            MI      0.07
         2. SUB-MEDITERRANEAN NEEDLE TREES                 MI      0.01
       + 3. MEDITERRANEAN EVERGREEN SHRUBS                 MI      0.48
         4. PALMIFORM TUFT-TREELETS                        PMTMAX  0.02
         5. DWARF-NEEDLE SMALL TREES                       PMTMAX  0.02
       + 6. SHORT BUNCH-GRASSES                            TMAX    0.68
         7. NEEDLE-LEAVED EVERGREEN SHRUBS                 TMAX    0.48
       + 8. MEDITERRANEAN DWARF-SHRUBS                     TMAX    0.37
         9. Various succulents and forbs

CEDRES                         AL-LUBNAN (Lebanon)             1930m
 17.5   0.5  768.  190.    0.    7.    1.35
       * 1. SUB-MEDITERRANEAN NEEDLE TREES                 TMIN    0.09
         2. NEEDLE-LEAVED EVERGREEN SHRUBS                 TMAX    0.38
         3. SHORT GRASSES                                  TMIN    0.67
         4. SUMMERGREEN FORBS                              TMAX    0.38

PERTH/WESTERN AUSTR.           AUSTRALIA                       65m
 24.0  13.0  883.  150.    8.   19.   1.14
       * 1. MEDITERRANEAN BROAD-EVERGREEN TREES            MI      0.30
       * 2. TROPICAL LINEAR-LEAVED TREES                   MI      0.12
         3. TROP. EG SCLEROPHYLL/MICROPHYLL TREES          TMIN
         4. SUB-MEDITERRANEAN NEEDLE TREES                 MI      0.25
         5. PALMIFORM TUFT-TREES & TREELETS                TMIN
         6. Broad-evergreen small trees                    MI
         7. BROAD-RAINGREEN SMALL TREES                    TMIN    0.23
         8. DWARF-NEEDLE SMALL TREES                       TMAX    0.50
         9. Mediterr. & Tropical BL-EG shrubs
        10. NEEDLE-LEAVED EVERGREEN SHRUBS                 TMAX    0.44
        11. Various tall and short grasses
        12. Various succulents and forbs
        13. MEDITERRANEAN DWARF-SHRUBS                     MI      0.06

SAVANNAH/GEORGIA               U.S.A                           15m
 27.5  10.4 1162.  180.   40.  180.   1.20
       * 1. WARM-TEMPERATE BROAD-EVERGREEN TREES           TMAX    0.25
       * 2. SUMMERGREEN BROAD-LEAVED TREES                 TMAX    0.17
       * 3. HELIOPHILIC LONG-NEEDLED TREES                 MI      0.17
         4. Trop. EG sclerophyll/microphyll trees          TMIN
         5. Many small trees, shrubs, and other forms
```

Table 1. Vegetation profiles for selected sites with aromatic vegetation.

Vegetation profiles are shown for three sites with typical Mediterranean vegetation and for two sites outside the Mediterranean region which also have summer water stress and emitting vegetation. The situation at Iraklion on Crete is most typical of potentially harvestable mediterranean vegetation, showing both evergreen sclerophyll shrubs and aromatic smaller shrubs (for classification of plant types see Table 2). Tarifa in southern Spain represents a coastal mediterranean forest (open), and Cèdres represents the montane cedar forests of Lebanon.
Perth, on the coast of southwestern Australia (32°S), is also in a mediterranean climate but shows a strong admixture of tropical forms, since summer dryness is not so extreme (8 mm in the driest month and 19 mm in the warmest). Savannah, near the coast in Georgia (southeastern USA, 32°N), has its rainfall maximum in the summer. Between heavy showers, however, and because the soil is very sandy, water stress can occur daily and can continue for weeks if rainfall is irregular. The main emitting plants in this region are the sclerophyllous "southern pines" (heliophilic long-needled trees) plus deciduous and evergreen oaks.
The vegetation profiles are generated by means of climatic envelopes (as in Table 2) for each of 90 basic types (4) plus a generating program ECOSIEVE. The relative importance of the different forms in a given situation is suggested by the numbers in the last column, which represent (standardized) distance to the closest climatic limit (preceding column). Small numbers indicate forms occurring near climatic limits. Missing information in these colums and/or any departure from totally capitalized names indicates that some forms have been grouped in order to save space. Some less important forms have been omitted. Comparison at 74 validation sites of actual vegetation and the predicted profiles suggested an accuracy of about 80%, with most problems in areas with edaphic problems. Asterisks and/or plusses before plant names indicate predicted (co-)dominant forms.

BL = broad-leaved EG = evergreen

	TMAX	TMIN	PRCP	MI	PMAX	PMIN	PMTMAX
Mediterranean BL-EG Trees	32	15	---	2.5	--	75	75
(e.g. *Quercus, Olea, Arbutus*)	20	5	500	0.8	60	0	0
Sub-mediterranean Needle Trees	28	15	---	3.0	--	100	100
(e.g. *Cedrus, Pinus, Cupressus*)	14	-1	400	0.85	75	0	0
Dwarf-Needled Small Trees	30	18	---	3.0	300	100	100
(e.g. *Juniperus, Tamarix*)	18	-5	200	0.45	25	0	5
Mediterranean EG Shrubs	30	18	---	2.0	--	100	125
(e.g. *Myrtus, Quercus*)	16	5	300	0.45	60	0	0
Needle-Leaved EG Shrubs	35	15	---	2.0	300	100	300
(e.g. *Juniperus, Rosmarinus*)	10	-30	100	0.4	40	0	0
Mediterranean Dwarf-Shrubs	35	15	---	1.2	--	50	50
(e.g. *Thymus, Salvia, Cistus*)	16	3	150	0.15	35	0	0
Summergreen Forbs	35	15	---	---	--	--	--
(e.g. *Lamium, Mentha*, geophytes)	10	-50	100	0.2	30	0	0

Broad-Evergreen Sclerophyll Trees (e.g. *Eucalyptus*)

Tropical Xeric Needle-Trees (e.g. *Juniperus*)

Heliophilic Long-Needle Trees (e.g. *Pinus caribbea, P. taeda*)

Temperate Needle-Trees (e.g. *Pinus*)

Hot-Desert Evergreen Shrubs (e.g. *Larrea tridentata*)

Cold-Winter Xeric Shrubs (e.g. *Artemisia*)

Xeric Summergreen Shrubs (e.g. "deciduous chapparal")

Table 2. Climatic envelopes of mediterranean aromatic plant forms.

The table presents the main mediterranean plant life forms, according to the system of Box (4), and their climatic relations, plus a listing of some other basic ecological plant types which are also known for emission of aromatic or other secondary substances. The climatic variables (left to right across the top) are mean temperature of the warmest and coldest months, average annual precipitation, a moisture index defined by PRCP divided by annual potential evapotranspiration (Thornthwaite estimate), and the average precipitation amounts for the wettest, driest, and warmest months. Temperature is in ^{o}C and precipitation amounts in millimeters; dashes represent undefined (infinite, presumably unimportant or unattained) limits. Aromatic shrubs are mainly the Mediterranean Dwarf-Shrubs. The evergreen sclerophyll vegetation of maquis includes the Mediterranean EG Shrubs, Needle-Leaved EG Shrubs, and Dwarf-Needled Small Trees. The two tree forms are generally restricted to more moist coastal and montane mediterranean climates.

BL = broad-leaved NL = needle-leaved EG = evergreen

These climatic envelopes were used to generate the predicted vegetation profiles shown in Table 1.

under varying environmental conditions. The production, respiration, and phytomass accumulation/partitioning processes of vegetation in general are currently being studied at the world scale with such a model (PHYTMASS: Box, unpublished data). This model contains deterministic functions for gross production (GPP) and respiration (Rd) as follow:

$$GPP = GPP_{max}\ (1 - e^{-a \cdot AT}) \qquad (2)$$

$$Rd = B \cdot R_o \cdot e^{b(T-T_o)} \qquad (3)$$

where GPP_{max} is an upper limit for GPP (tropical rainforest); AT is actual transpiration (dependent on current leaf phytomass but never exceeding AET); B is the current total phytomass; T is the current temperature; R_o and T_o represent a reference point for respiration and temperature; and a and b are empirical coefficients. Net production (NPP) and phytomass increment (ΔB) are determined from the general material-balance equations:

$$NPP = GPP - Rd \qquad (4)$$

$$\Delta B = NPP - LP \qquad (5)$$

where LP (litter production) is derived from an annual value at climatic climax given by:

$$LP = 1.0066\ AET - 183.92\ (g\ m^{-2}\ year^{-1}) \qquad (6)$$

Such simulation models can also be coupled with computer cartographic systems and large data-bases to provide output as predictive maps. Geographic predictions of this sort are especially useful for initial evaluation of models since errors are usually quite visible on the maps. An initial predictive map of equilibrium phytomass accumulation on the world's land areas has been produced (Box, unpublished data) and appears reasonable in most areas.

For Iraklion (see Table 1), a relatively moist, coastal Mediterranean site, both PHYTMASS (Box, unpublished data) and equation (1) estimate total annual net productivity to be around 1100 g m^{-2}. This compares favorably with the 600 g $m^{-2}year^{-1}$ above ground which Margaris (10) suggests as typical for Greek maquis-type vegetation. Equation (6) suggests an annual litter production of around 300 g m^{-2}, leaving an annual increment approaching 800 g m^{-2}. If this amount can be attained each year, with perhaps 75% above ground for vegetation resprouting after fire, then it would take about 10 years to attain the 6 kg m^{-2} which Margaris suggests as typical for above ground phytomass in Greek maquis. This quick recovery, relative to many other ecosystems, appears to be due to the relatively low amount of standing phytomass climatically attainable and the fact that a large fraction of the post-fire production is directed into above-ground parts. The total amount of litter produced during this time would be about 3 kg m^{-2}, of which around half (Meentemeyer, personal communication) might have accumulated without decomposition. The potential for fire during these 10 years is, of course, a function of weather, the amount of lying litter, and various human factors.

4. PRODUCTIVITY MODELLING OF PLANTS AND PLANT TYPES

Due to different structures and physiologies, different plants have different water and energy budgets, different climatic limits, and require different climatic coefficients for modelling. Although each species may be different, the life-form system described earlier provides a means of simplifying this potentially infinite variation. One can imagine equations similar to (2)-(6) for each of the life forms in Table 2, with coefficients related to structural characters such as leaf area and "hardness". The result might contribute to what can be called life-form ecology, the eco-physiognomically based ecology of groups of plants with similar require-ments due to similar form and thus similar water and energy budgets. Of course, shading and other interactions between forms would play a role in any natural stand.

The logic of this sort of life-form productivity modelling can perhaps best be seen from Fig. 1. As transpiration (AT) increases (temperature also increasing), one normally expects gross production to increase toward some upper limit (saturation curve, as in equation 2) and respiration to

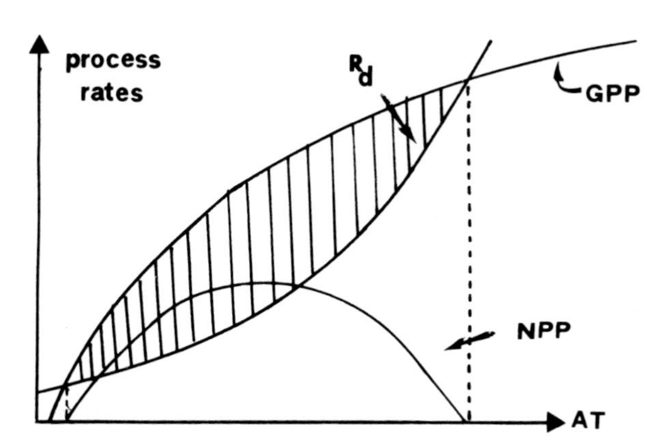

FIGURE 1. Logic of ecophysiognomic functional modelling: typical relations of gross and net production and respiration.

increase, on the other hand, at an increasing rate (exponential curve, as in equation 3) with Q_{10} value around two. As long as gross production exceeds respiration, net production can be conceived to be the difference between the two curves and will describe a parabola-like curve on the same axes. A formula for net production could be derived, then, either from the difference between equations (2) and (3):

$$NPP = GPP_{max} \left(1 - e^{-a \cdot AT}\right) - B \cdot R_o \cdot e^{b(T-T_o)} \qquad (7)$$

or, more conveniently, from the general parabolic equation:

$$NPP = NPP_{max} - \mu \left(AT - AT_o\right)^2 \qquad (8)$$

where NPP_{max} is the maximum value, AT_o is midway between the zero-points, and μ depends on NPP_{max} and AT_o. In either case, the curve will have two zero-points along the AT-axis. If these cardinal values can be derived from the climatic limits of Table 2, then a means of parameterizing such models is provided. The upper and lower limits for TMAX (mean temperature of warmest month) might serve as the best initial approximation, since they generally refer to the growing season.

5. MODELLING EMISSIONS PRODUCTIVITY

One aspect of productivity which has not been studied so much but which is important to the study of aromatic vegetation is the question of how much photosynthate is directed into the production of the secondary substances. Zimmerman (14) has collected many data on emissions of terpenes and isoprene from emitting taxa of various, mainly temperate North American climates. These data, however, have not yet led to formulation of general models, mainly because the values from field measurements vary so much, due to both genetic and microclimatic variability.

The following can be said so far about such emissions:

- They involve substances stored and probably produced in leaves.
- Some emissions show light-sensitivity while others do not.
- Emissions generally increase with temperature.
- Emissions often appear to be correlated more with leaf area/ phytomass than with gross or net production rates.
- Emissions occur especially after rain.
- Measured emission rates in the field vary enormously, both from species to species and within species.

In connection with the apparent correlation with leaf phytomass, Box (5) has attempted an overview of leaf-phytomass data. Emission of secondary substances depends, however, on microclimatic conditions affecting the stomata. Total amounts of secondary substances produced should vary much less than emissions measured at particular times and places. Tingey (13 and personal communication) has outlined a program for studying the physiological behaviour of plant emissions when individual plants (eliminating genetic variation) are subjected to changing climatic conditions in the laboratory. Such studies should shed more light on ecophysiological mechanisms of secondary-substance production and emission.

Until more is known about determinants of secondary-substance production, modelling attempts must be considered preliminary. Nevertheless, the general logic of production processes plus the apparent relationship with water stress suggests that secondary substance production (SSP), annualy or for shorter periods, might be expressed by a saturation function of actual transpiration (AT, cf. equation 2) multiplied by some factor representing the tendency toward SSP in subhumid to arid climates. If that factor is conceived as a normal-distribution function centered on some appropriate value (MI_o) of an annual moisture index (MI), then one would have a model of the form:

$$SSP = SSP_{pot}\,(1 - e^{-c \cdot AT}) \qquad (9)$$

with SSP_{pot} being a normal-distribution function of MI given by:

$$SSP_{pot} = S_o \cdot N(MI_o, \sigma) = S_o \left[\frac{1}{\sigma\sqrt{2\pi}}\, e^{-\left(\frac{MI - MI_o}{\sigma}\right)^2 / 2} \right] \qquad (10)$$

where σ is standard deviation, and c and S_o are empirical coefficients. Note that equation (9) has the same general form as equation (2) for gross production but with SSP_{pot} always much less than total production. Whether an additional coefficient for taxonomic differences is necessary in equation (9) must still be studied.

This formulation conceives SSP as a function of actual transpiration and of climate-based allocation strategies. The more readily measurable emission of such secondary substances remains a more ephemeral phenomenon but should not exceed SSP in amount, thus providing a means of estimating coefficients c and S_o.

6. CONCLUSIONS

Various studies have now shown that well-chosen macroclimatic variables, especially combinations of temperature and water availability, are reliable predictors of vegetation processes, including net and gross production, dark respiration, and litter production and decomposition. Structural aspects of vegetation (e.g. leaf area, litter and phytomass accumulations) represent balances among several processes and are more complex but also show discernible relations to climatic determinants. Climate-based models can be especially useful for studying year-to-year variability. The success of the life-form approach in predicting world vegetation patterns suggests that ecophysiognomic considerations are a useful basis for modeling plant-environment relations. Estimation of the amount of energy directed into the production of secondary substances is problematic, due to the ephemeral nature of emissions, but can probably be done once determinants are better understood.

The question of harvesting mediterranean vegetation involves not only questions of how much but also how often. This latter aspect involves understanding of the entire vegetation recovery process, including nutrient loss and litter accumulation as well as the accumulation of usable phytomass.

The apparent relationship between emitting plants and summer water stress suggests the geographic question: where else can economically useful aromatic plants be grown? The most obvious general region appears to be the subtropical east-coast climates, especially of North America, China, and Australia, where clayey soils are similar and soil dryness in summer produces very mediterranean-like water availability conditions except in unusually wet years. Such climates, however, are generally more continental than the mediterranean and have lower winter temperature extremes.

REFERENCES

1. Box EO. 1975. In: Primary Productivity of the Biosphere, Lieth H and Whittaker RH (eds.). New York, Springer.
2. Box EO. 1978. *Radiat. & Envl. Biophys*, 15:305
3. Box EO. 1979. Quantitative Cartographic Analysis: A Summary (with Geoenvironmental Applications) of SYMAP Auxiliary Programs Developed at the Jülich Nuclear Research Center. Jülich: Reports of the Kernforschungsanlange Jülich GmbH, no. 1582.
4. Box EO. 1981. In: Macroclimate and Plant Forms: An Introduction to Predictive Modeling in Phytogeography. Tasks for Vegetation Science, vol. 1. Den Haag, Dr. W. Junk BV.

5. Box EO. 1981a. In: Proceedings of EPA Symposium on Atmospheric Biogenic Hydrocarbons. (See ref. 6).
6. Environmental Protection Agency. 1981. Proceedings of the Symposium on Atmospheric Biogenic Hydrocarbons. Research Triangle Park, North Carolina, January 1980, Arnts R (ed.). Ann Arbor/Michigan, Ann Arbor Science Series.
7. Lieth H, Box EO. 1972. *Climatology* 25:37
8. Lieth H, Box EO. 1977. *Tropical Ecology* 18:109.
9. Major J. 1963. *Ecology* 44:485.
10. Margaris NS. 1979. In: Biological and Sociological Basis for a Rational Use of Forest Resources for Energy and Organics, Boyce SG (ed.) p.121. USDA/Forest Service, Southeast. For. Exp. Stn. Asheville, N.C.
11. Meentemeyer V. 1978. *Ecology* 59:465
12. Rosenzweig LM. 1968. *Amer. Naturalist* 102:67.
13. Tingey DT. 1981. In: Proceedings of EPA Symposium. (See ref. 6).
14. Zimmerman PR. 1979. Testing of Hydrocarbon Emissions from Vegetation, Leaf Litter, and Aquatic Surfaces and Development of a Methodology for Compiling Biogenic Emissions Inventories. Report no. EPA-450/4-79-004. Research Triangle Park/North Carolina, US Environmental Protection Agency.

A TAXONOMIC REVISION OF *SIDERITIS* L. SECTION EMPEDOCLIA (RAFIN.) BENTHAM (LABIATAE) IN GREECE

K. PAPANIKOLAOU, S. KOKKINI

1. INTRODUCTION

The genus *Sideritis* section Empedoclia has long been known as part of the Greek flora. In Hayek's (11) flora ten taxa were accepted for Greece, viz. *S. clandestina* (Chaub. & Bory) Hayek, *S. clandestina* (Chaub. & Bory) Hayek var. *cyllenea* (Boiss.) Hayek, *S. scardica* Gris., *S. scardica* Gris. var. *pelia* Hal., *S. perfoliata* L., *S. raeseri* Boiss. & Heldr., *S. raeseri* Boiss. & Heldr. var. *lanceolata* Hal., *S. raeseri* Boiss. & Heldr. var. *attica* (Heldr.) Hal., *S. euboea* Heldr., and *S. syriaca* L. Heywood (12) reduced the number of taxa occurring in Greece to four, viz. *S. syriaca* and *S. clandestina* (Chaub. & Bory) Hayek, *S. scardica* Gris., and *S. perfoliata* L.

No comprehensive taxonomic treatment of *Sideritis* section Empedoclia has been published since the revision by Hayek in Prodromus Florae Peninsulae Balcanicae.

Within the *S. raeseri* complex it is possible to recognize a number of regional form series which may be of interest for the future study of the history of the mountain flora. The pausity of clear-cut morphological discontinuities as well as the pronounced local differentiation excludes the use of specific status for these form series. The complex has therefore been treated as a single species, subdivided into three allopatric, reasonably distinct subspecies.

This paper gives an up-to-date taxonomic and nomenclatural revision of the section, resulting in the recognition of three species and seven subspecies. Morphological aspects are briefly reviewed.

2. PROCEDURE

2.1. Material and methods

This revision is mainly based on herbarium studies and observations made in the field. Five taxa have also been kept in cultivation. Material from the following herbaria has been studied (abbreviations according to Lanjouw and Stafleu (14)): C, G, Herb. Thessaloniki, K, LD, W, and B. Aldén's, L.Å.

Margaris N, Koedam A, and Vokou D (eds.): Aromatic Plants: Basic and Applied Aspects

ISBN 90-247-2720-0. Printed in the Netherlands.

Gustavsson's, P. Hartvig's, A. Strid's private collections. All material available (ca 600 sheets) has been determined and labelled.

Morphological observations and measurements were conducted under a Leitz stereo-microscope. Leaves as well as bracts have been softened by soaking in hot water and flattened between two slides, whilst calyces and corollae were studied in the form of dry herbarium material. The descriptive terminology is generally in accordance with Stearn (19). The descriptions normally attempt to cover the total range of variation. Rare extreme values are sometimes given in brackets.

The drawings have been made from dried material (floral parts and bracts have been softened by soaking in hot water).

Specimens without locality or having only very vague locality statements have been omitted.

2.2. Taxonomic concept

The taxonomic concepts applied in this study do not differ from those commonly adopted in contemporary taxonomic revisions based mainly on herbarium material. Two populations have thus been distinguished as separate species if they differ by a combination of more or less sharp discontinuities in several morphological characters. Subspecies rank has been applied to geographically and/or ecologically allopatric groups of populations which are morphologically distinguishable, but may merge into each other in interadjacent zones of limited extension. Even though field observations on ecology, variation, flowering period etc., have been given due consideration, it is evident that the subdivision presented here is by no means the final answer to all taxonomic problems in the section Empedoclia. Especially at subspecies level the taxonomy is necessarily subjective and preliminary to a certain extent. Variety rank, which is often somewhat ambiguous, has also been avoided.

2.3. Chromosome numbers

The genus *Sideritis* section Empedoclia is not suitable for cytological studies. Seeds of five taxa were sown in the experimental field of the Copenhagen University Botanical Garden and chromosomes were counted in root-tip preparations fixed in Navashin-Karpechenko, cut by means of a microtome, and stained with crystal violet. The chromosome number 2n=32 was observed in all taxa studied. Contandriopoulos (7) has reported the same chromosome number for all taxa of the section Empedoclia.

2.4. Ecological aspects

Weimarck's (23) phytogeographical groups were based exclusively on distribution patterns of members of typical Cape genera. Strid (20) has pointed out that ecological specialization is also a distinctive feature of many Cape plants, among them several species of *Adenandra*. Dahlgren (8) has demonstrated frequent correlations between substrate and distribution patterns is *Aspalathus*. Nevertheless, all taxa of the genus *Sideritis* section Empedoclia grow on limestone background. They are confined to stony places in alpine grasslands, with the exception of one population of *S. athoa* which grows in rock crevices of hard limestone on Mt Pachtourion (S Pindhos) (cf. Aldén (1)), and *S. raeseri* ssp. *attica* which grows in rock crevices, too. They do not occur on schistose mountains. In labels of some herbarium specimens "substrate schist" is written, but in the mountains where they come from both schist and limestone occur (i.e. Mt Tzena).

2.5. Vicarism

Some pairs of groups of closely related taxa are geographically vicarious. These are the following:

1. *S. clandestina* - *athoa*. The former is endemic to Peloponnese while the latter occurs on Mt Athos, island of Samothrace, S Pindhos and Mt Ida (W Anatolia). The above two species are the only ones with two stripes on the two lobes of the upper corolla lip. *S. scardica* is something between *S. raeseri* and *S. clandestina* in all resepcts.
2. *S. syriaca* - *raeseri* - *euboea*. The first species is endemic to Cretan mountains while the second and third occur in Sterea Hellas mountains, Pindhos range, and Euboea mountains, respectively.
3. Subspecies of *S. raeseri*. All subspecies of *S. raeseri* are allopatric. Ssp. *attica* occurs in Mts Parnis, Pateras and Kitheron, ssp. *raeseri* from Mt Parnassos northwards extending into Jugoslavia as well as into Albania; ssp. *florida* is endemic to Mt Olympos in NC Greece.

Further comments on distribution and affinities are found under the respective species and subspecies entries.

3. TAXONOMY

3.1. Key to species

1. Middle and upper stem leaves rounded or cordate at base, semiamplexicaul and often perfoliate..1. *S. athoa*

Middle and upper stem leaves attenuate at base, not semiamplexicaul and never perfoliate...2

2. Corolla with two stripes on the lobes of the upper lip; basal leaves oblong-spathulate, obtuse; a gradual transition from stem leaves to bracts; acumen 5(ssp. *cyllenea*) - 15(ssp. *clandestina*) mm long...2. *S. clandestina*

Corolla without two stripes on the two lobes of the upper lip; basal leaves oblong-lanceolate or lanceolate, subobtuse; no gradual transition from stem leaves to bracts; acumen 2-7 mm long................ 3

3. Middle bracts exceeding flowers, 16-23 X 14-19 mm; acumen 4-7 (10) mm long; verticillasters crowded to form a dense spike........ 3. *S. scardica*

Middle bracts shorter or equalling flowers, 6-25 X 6-22 mm; acumen 2-4 mm long... 4

4. Bracts glandulous-puberulent outside (glands rather long-stalked)..... .... 4. *S. raeseri*

Bracts densely white-tomentose outside.................... 5

5. Verticillasters distant; plant 30-50 cm tall............ 5. *S. syriaca*

Verticillasters compact; plant 15-35 cm tall............. 6. *S. euboea*

1. *Sideritis athoa* Papanicolaou & Kokkini sp. nov. (Figs. 1A, 1B, 11A, 12A)

Orig. coll.: NE Greece: Chalkidiki peninsula. Mt Athos, SW side, above Panaghia, 1700 m Papanicolaou no.229, 17.7.1978 (holotype in Herb. Thessaloniki, isotypes in C, LD).

Syn.: *S. perfoliata* L., Grisebach (9) in Spic. Flor. Rum. 1: 141 (1843); Boissier (4) in Flor. Orient. 4: 714 (1879) pro parte; Rechinger (17) in Fl. Aegaea (1843) pro parte; Heywood (12) in Fl. Eur. 3: 142 (1972); Aldén (1) in Bot. Not. 129: 302 (1976).

Planta basi suffrutescens. Caules erecti vel ascendentes, 13-30(40) cm alti, simplices vel 1-ramosi, appresse albo-tomentosi vel glabrescentes. Folia caulina media lanceolata-oblonga, 3-5 cm longa, 1.2-1.5 cm lata, acuta, crenato-dentata vel integerima, reticulato-venosa, semiamplexicaulia sessilia, panosa vel glabrescentia. Verticillastri inferiores remoti superiores approximati. Bracteae mediae late ovato-lanceolata, acuminata. Calyx

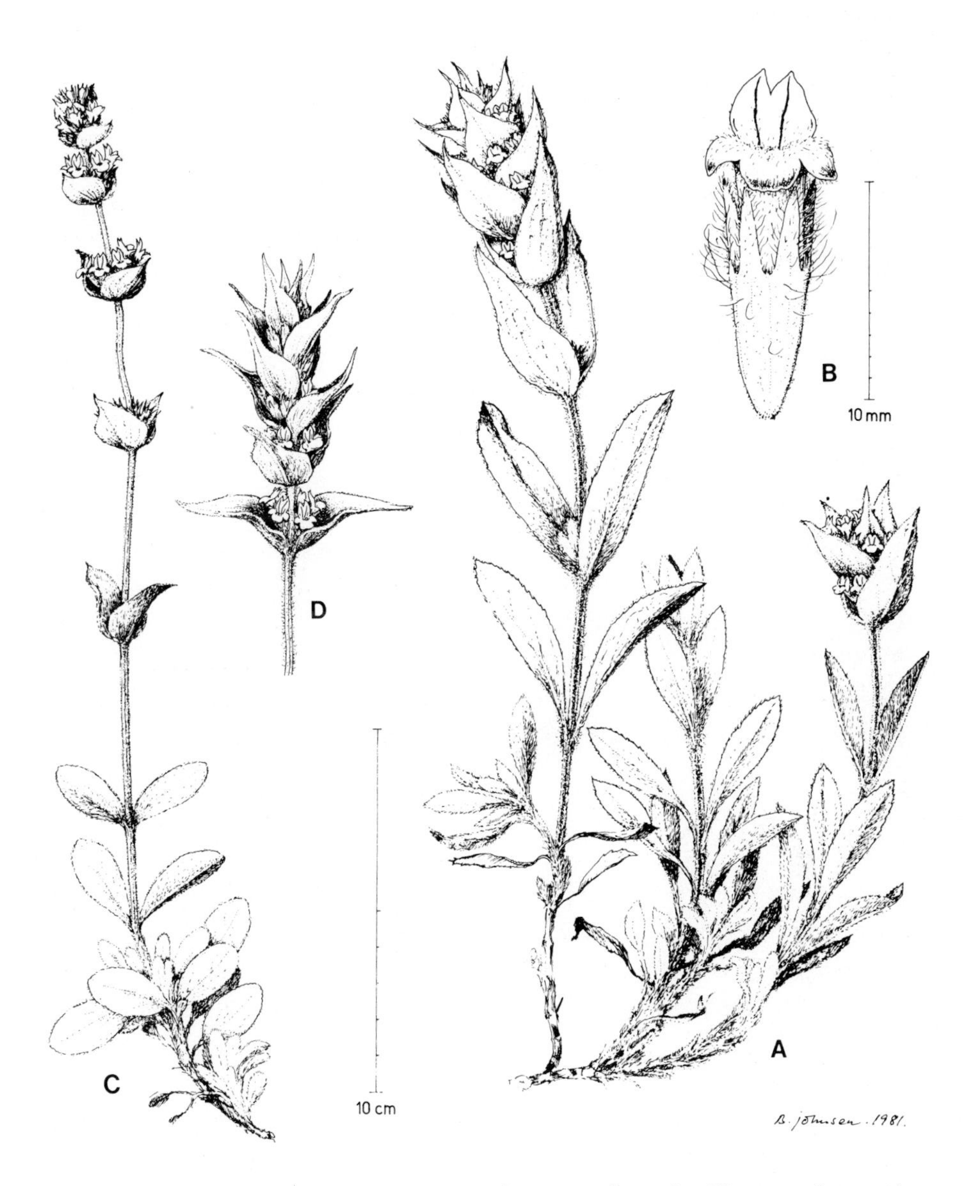

FIGURE 1. A: Habit of *Sideritis athoa* (Holotype). - B: Flower of *S. athoa*. C: Habit of *S. clandestina* ssp. *cyllenea* (Lectotype). - D: Inflorescence of *S. clandestina* ssp. *clandestina* (Lectotype) for comparison.

FIGURE 2. *Sideritis perfoliata*. Photo of the Linnaean type specimen.

10-11 mm longus, pilis eglandulosis 4 mm longis, dentes 4-4.5 mm longi, acuti vel acuminati. Corolla flava, 12-13.5 mm longa, sparse glandulosa.

Perennial, densely white-tomentose or glabrescent, with tiny glandular hairs all over herb. Base of stem more or less woody, producing several erect or ascending, unbranched or 1-branched above flowering shoots, 13-30 (40) cm long. Basal leaves light green or green-yellowish, coriaceous, lanceolate or oblong-lanceolate, acute, usually dentate, attenuate at base, 40-60 X 7-12 mm (including petiole). Middle and upper stem leaves coriaceous, strongly reticulate, rounded or cordate at base, lanceolate or oblong-lanceolate, semiamplexicaul, perfoliate, 30-50 X 9-16 mm. Inflorescence usually unbranched, 5-22 cm long. Verticillasters 4-10, compact below, more or less distant above. Bracts entire, ovate-triangular, 2-3 times as long as flowers, acuminate 25-35 X 15-22 mm with an acumen 10-15 mm long, becoming shorter and narrower towards the apex of inflorescence. Calyx campanulate, glandular all over, with eglandular hairs only towards the teeth ca 4 mm long, 10.5-12 mm long, with teeth 4-5 mm long. Corolla yellow with two stripes on the two lobes of the upper lip, 12-14.5 mm long.

Note: Our new species had apparently been identified as *S. perfoliata* L. by Grisebach (9) on material collected by Friedrichsthal and preserved at Vienna. This mistake was repeated subsequently by other authors (Boissier (4), Hayek (11), Rechinger (17), Heywood (12)). When the plant was recollected and cultivated (1978) it became clear that it is totally distinct from *S. perfoliata*, which is certainly an Asiatic species without specific locality. At Kew there is a specimen collected by Adamović (VII, 1903) named by the collector as *Sideritis athoa* Adamov., which remains apparently an unpublished manuscript name. Adamović had perhaps seen the Linnaean type specimen.

The Athos material does not fit the Linnaean type specimen seen in microfiche (Fig. 2). The citation of *S. perfoliata* in Species Plantarum (15) is: "Habitat in Oriente?". Asiatic specimens, without specific citation (just "Asia") seen by the authors seem to be identical with the picture of the Linnaean plant in microfiche. The material from Mt Ida (W Anatolia), the island of Samothrace and S Pindhos is similar to the Athos plants. Turrill (22) has mentioned that "the Athos specimens at Kew are nearly as lanate as the Samothrace specimens and the study of more material from both localities might enable the north Aegean plants to be more definitely separated from those of Asia Minor as one variety".

Relationships: *Sideritis athoa* is undoubtedly related to *S. perfoliata* s.str. from Asia as well as to *S. clandestina*, endemic to Peloponnese, but differs in a number of characters (Table 1). The two stripes on the two lobes of the upper corolla-lip (the rest Greek taxa have no such stripes) as well as the bract shape and size are the most important common characters between the latter species and *S. athoa*.

Flowering period: From beginning of July to mid-August.

Distribution: *S. athoa* is the only Greek taxon of the section Empedoclia with such a discontinuous distribution (Fig. 3). It occurs on Mt Ida (W Anatolia), island of Samothrace, Mt Athos and Mt Pachtourion in S Pindhos. In Pindhos, it is also reported from Mt Agrafa (Heldreich in Halácsy (10)), Mt Neraidha (16). Specimens (not seen by us and perhaps wrongly determined) have been referred to as *S. perfoliata* from Mt Grammos (24) and Mt Avgho (18).

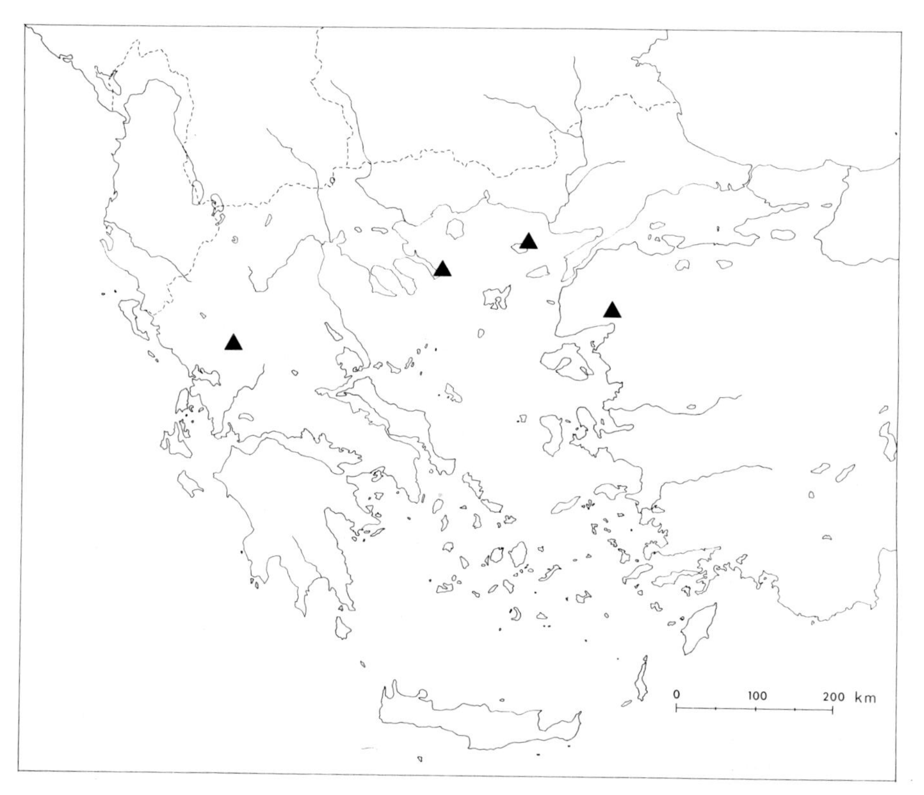

FIGURE 3. *Sideritis athoa*. Distributional map based on material checked by the authors.

Table 1. Differences between *Sideritis athoa, S. perfoliata* and *S. clandestina*. Measurements based on herbarium material checked by the authors.

S. athoa	*S. perfoliata*	*S. clandestina*
Plant 13-30(40) cm tall	Plant 25-60 cm tall	Plant 18-30(38) cm tall
Basal leaves lanceolate, attenuate at base, shortly petiolate, 40-60 X 7-12 mm	Basal leaves ovoid or ovoid-oblong, shortly petiolate, 60-90 X 15--22 mm	Basal leaves oblong-spathulate, rather abruptly contracted into a long petiole, 50-80 X 14-20 mm
Upper stem leaves rounded or cordate at base, lanceolate or oblong-lanceolate, acute, perfoliate, 30-50 X 9-16 mm	Upper stem leaves cordate at base ovate-oblong, acute, perfoliate, 60-80 X 17-25 mm	Upper stem leaves subsessile, rounded to attenuate at base, lanceolate or oblong-lanceolate, subacute , 30-60 X 10-18 mm
Inflorescence 0-1-branched	Inflorescence 1-4--branched	Inflorescence unbranched
Bracts ovate-triangular, broadly acuminate, becoming shorter and narrower towards the apex of inflorescence	Bracts ovate or ovate-reniform, rather abruptly contracted into an acumen, becoming shorter but wider towards the apex	Bracts ovate-triangular, rather abruptly contracted into an acumen of 5-15 mm long
Calyx 8-11.5 mm long, with eglandular hairs only towards teeth	Calyx 13-14.5 mm long, without or sparely eglandular hairs on tube (Fig. 11B)	Calyx 7-11(13) mm long, with eglandular hairs all over

Habitat: *S. athoa* does not appear to be very selective in the choice of locality, e.g. in Pindhos it was found by Aldén (1) to grow in rock crevices of hard limestone, while on Mt Athos it occurs in gravelly stony and rocky places with non-closed plant communities.

Variation: Plants from Pindhos are short and sparsely hirsute but such forms occur on Mt Athos too. Furthermore, some individuals from Mt Athos are as lanate as the Samothrace plants. Concerning the bract size there is a gradual increase from Pindhos to Mt Ida (Fig. 12A). No intermediate forms have been found between *S. athoa* and *S. perfoliata*.

Collections: Mt Athos. Sintenis & Bornmüller 945, 2.7.1891 (FI, G, K, LD, W); Janka 26.7.1871 (FI, W); Adamović, July, 1903 (K); Papanicolaou 229, 17.7.1978 (C, H.Th.); Dimonie, June, 1908 (FI). - Mt Ida. Sintenis 1179, 27.7.1883 (LD). - Island of Samothrace, Mt Fengari. Rechinger 9879, 18.6.1936 (G, K, LD, K). - Mt Pachtourion. Aldén 4707, 30.7.1974 (LD).

2. *Sideritis clandestina* (Chaub. & Bory) Hayek, Chaubard et Bory in Expéd. scient. de Morée, 170, t. 20 (1832).

Type: Des hautes régions du Taygète, où il nous a paru rare et d'où l'a rapporté M. Virlet lors de sa dernière ascension par les pentes orientales (G lectotype, selected here).

Basionym: *Phlomis clandestina* Chaub. & Bory, Expéd. scient. de Morée, 170, t. 20 (1832).

Syn.: *S. theezans* Boiss. & Heldr., Diagn. Pl. Orient. Nov. Ser. 1, 7: 58 (1846); *S. cretica* Sibth. & Sm., Prodr. 1: 40 pro parte, non L.; *S. syriaca* Chaub. & Bory, Nouv. flor. Pelop. (1838), non L.; *S. peloponnesiaca* Boiss. & Heldr., Diagn. Pl. Orient. Nov. Ser. 2, 3: 32 (1859).

Nomenclatural notes: *S. clandestina* was first described by Chaubard & Bory (6) under the name *Phlomis clandestina*. The original material came from "des hautes régions du Taygète, où il nous a paru rare, et d'où l'a rapporté M. Virlet lors de sa dernière ascension par les pentes orientales". The plate is reproduced in their (5) Flor. Pelopon. with the name *Sideritis syriaca* L. but this is not the Linnaean *S. syriaca*, which is a Cretan endemic plant. Boissier (2) and Heldreich published the name *Sideritis theezans* with a description of the same species. He (3) also published the name *S. peloponnesiaca* as a species. Hayek (11) has regarded it as a variety of *S. clandestina*, but in this paper it is treated as a subspecies of *S. clandestina* too.

Description: Perennial, grey-or yellowish-lanate herb. Base of stem more or less woody, producing several ascending, unbranched or branched above flowering shoots 18-30(38) cm long. Basal leaves moderate thick, densely tomentose, crenate or entire, oblong-spathulate, obtuse rather abruptly contracted into a long petiole, 50-80 X 14-20 mm (including petiole). Upper stem leaves subsessile, rounded to attenuate at base, lanceolate or oblong-lanceolate, subacute, 30-60 X 10-18 mm. Inflorescence usually unbranched. Verticillasters compact or distant. Bracts ovate-triangular rather abruptly contracted into an acumen of 5-15 mm long, 2-3 times as long as flowers, becoming shorter and narrower towards the apex of inflorescence, 11-33 X 6-20 mm. Calyx dense-hairy all over with hairs 1-3 mm long, narrowly campanulate, 7-11(13) mm long, with teeth 3-5 mm long. Corolla yellow with two stripes on the two lobes of the upper lip, 12-14 mm long.

Flowering period: From beginning of July to mid-August.

Distribution: *S. clandestina* is endemic on mountains of Peloponnese (Fig. 13).

Habitat: It is a plant of alpine and subalpine zones growing in stony places in grasslands.

Variation: The species has been divided in two subspecies. The most widespread of these is ssp. *cyllenea*, which occurs in northern Peloponnese mountains (Figs. 4, 13), characterized by a long inflorescence with distant verticillasters, smaller and sparsely lanate bracts with an acumen 4-8 mm long, and a habit somewhat between ssp. *clandestina* and *S. raeseri* ssp. *raeseri*. Ssp. *clandestina* is restricted to Mt Taygetos and Mt Parnon (Malevo Oros). It has a shorter inflorescence with compact verticillasters, larger (Fig. 1D), densely lanate bracts with an acumen 9-15 mm long (Figs. 4, 12C).

3.1.1. Key to the subspecies

1. Verticillasters compact. Bracts densely lanate, 16-33 X 10-20 mm, with an acumen 6-15 mm long; stem leaves equal or longer than internodes...... .... ssp. *clandestina*
2. Verticillasters distant. Bracts tomentose, 11-20 X 10-17 mm with an acumen 4-8 mm long; stem leaves shorter than internodes.... ssp. *cyllenea*

2a. *Sideritis clandestina* (Chaub. & Bory) Hayek ssp. *clandestina* (Figs. 1D, 4, 11C, 12C, 13).

Type collection: see under the species.

This subspecies is common on Mt Taygetos as well as on Mt Parnon, where

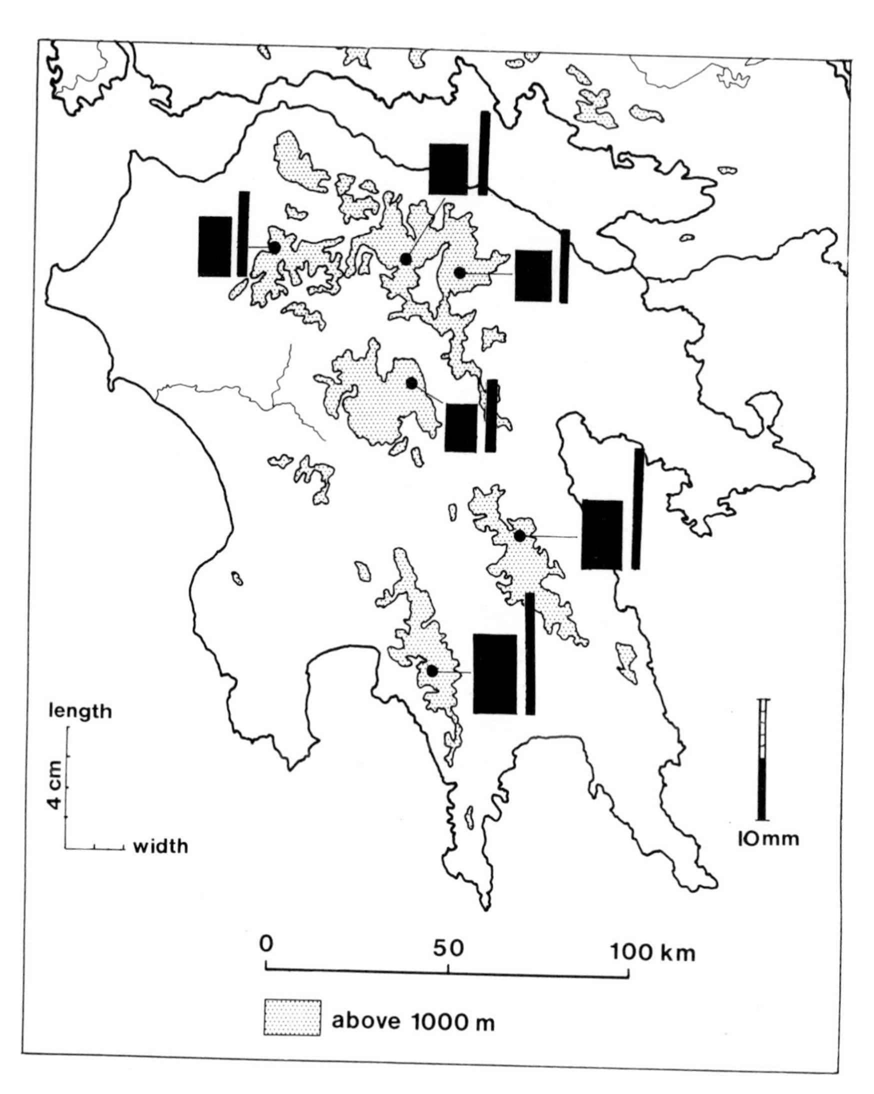

FIGURE 4. *Sideritis clandestina*. Histograms (to the left) showing the average mean values of the bract length (vertical axis) and width (horizontal axis) from its distribution area. The length (to the right) of acumen. The subdivision of the species is to some extent based on these characters. All measurements are taken from material collected in nature.

it had already been collected by Chaubard and Bory. Characteristic features include the compact inflorescence, long, densely lanate bracts, and stem leaves equal or longer than internodes.

Collections: Mt Taygetos. Strid & Papanicolaou 15269, 30.6.1979 (C); Pichler, 7.1876 (K, W); Leonis 199, 30.6.1902 (W); Regel, 5.8.1934 (G); Regel, 2.7.1935 (G); Heldreich, 1844 (G); Heldreich, 1848 (W); Baden & Franzén 631, 19.7.1980 (C); Barnaby, 9.1938 (K); Chaubard ?1830 (G). - Mt Parnon. Orphanides 14, 7.7.1850 (K, LD); Heldreich, 8.1844 (K); Heldreich 295 (K); Topazi, 14.7.1877 (K); Leonis 1367, 1.6.1896.

2b. *Sideritis clandestina* (Chaub. & Bory) Hayek ssp. *cyllenea* (Boiss.) Hayek) Papanicolaou & Kokkini comb. et stat. nov. (Figs. 1C, 4, 12D, 13).

Orig. coll.: In regione superiori montium Olenos et Kyllines Peloponnesi alt. 5500'-6500'. Heldreich, Jul.1848 (K lectotype selected here).

Basionym: *Sideritis peloponnesiaca* Boiss. & Heldr., Diagn. Plant. Orient. Nov. Ser. 2, 3: 32 (1859).

Syn.: *Sideritis theezans* Boiss. & Heldr. var. *cyllenea* Boiss., Flor. Orient 4: 711 (1879); *S. theezans* Boiss. & Heldr. var. *peloponnesiaca* (Boiss. & Heldr Hal.

This is a sparsely lanate plant with an inflorescence having verticillasters distant; bracts herbaceous, smaller than those of ssp. *clandestina* (Figs. 4, 12D); basal leaves ovate, to suborbicular, always smaller than those of ssp. *clandestina*, and stem leaves shorter than internodes. All the specimens seen have a habit somewhat between ssp. *clandestina* and *S. raeseri* ssp. *attica* (see below), which is reasonable as ssp. *cyllenea* occurs in the area between (Fig. 13).

Collections: Mt Olenos. Heldreich, 7.1848 (K). - Mt Chelmos. Baden & Franzén 849, 31.7.1980 (C). - Mt Killini. Strid & Papanicolaou 15382, 4.7. 1979 (C); Heldreich 25.6.1887 (LD). - Mt Menalon. Greuter 9381 (G).

3. *Sideritis scardica* Griseb., Spicil. Flor. Rumel. 2:144 (1844).

Type: In Scardo: Pratis montanis. Ljubatrim spartium in angusta regione alt. 3000'. Grisebach ? (K lectotype selected here, Göt).

Note: Hooker (13) has made an illustration of *S. scardica* using material from Mt Olympos. In this paper the Olympos plants have been treated as *S. raeseri* ssp. *florida* (see below).

Description: Perennial, suffruticose, densely white-tomentose herb.

Basal leaves moderate thick, oblong-lanceolate, usually crenate, subobtuse, mucronate, attenuate to a petiole, 40-80 X 7-10 mm. Middle and upper stem leaves subsessile, linear-lanceolate, acute, 30-60 X 6-8 mm. Inflorescence 3-8 cm long. Verticillasters very close each other. Bracts ovate, rather abruptly contracted into an acumen or acuminate, 1-2(2.5) times as long as flowers, becoming shorter and narrower towards the apex of inflorescence, 17-35(40) X 8-19 mm. Calyx dense-hairy all over, narrowly campanulate, 8-11 mm long; teeth 3-5 mm long. Corolla yellow without two stripes on the lobes of the upper lip, 12-14 mm long.

Flowering period: From beginning of July to mid-August.

Distribution: It occurs in CE, NC and NE Greece (Figs. 6, 13). Outside Greece in S Jugoslavia, S Bulgaria and SW Albania.

Habitat: It is a plant of alpine zone growing in stony places in alpine meadows.

Variation: The species has been divided in two subspecies. In Greece the most widespread is ssp. *longibracteata* (Figs. 6, 13), characterized by an inflorescence 5-8 cm long, bracts narrowly ovate to oblong with long, appressed eglandular hairs outside, 19-40 X 9-14 mm. Ssp. *scardica* in Greece is restricted to Mt Vermion and Mt Kajmakčalan extending northwards to Mt Scardo in S Jugoslavia. It has a shorter inflorescence (2.5-4 cm), and ovate bracts with more or less curled eglandular hairs outside, 17-22 X 15-19 mm (Figs. 5, 6, 12F). Some specimens from Mt Vermion have appressed eglandular hairs on bracts, also, from Mt Pangaion have a short inflorescence.

3.1.2. Key to the subspecies

1. Bracts broadly ovate, white-lanate, with curled, eglandular hairs outside, rather abruptly contracted into an acumen of 4-8 mm long, 17-22 X 15-19 mm .... ssp. *scardica*
2. Bracts narrowly ovate to oblong, herbaceous, with appressed eglandular hairs outside, 19-40 X 9-14 mm; acumen 5-20 mm long.. ssp. *longibracteata*

3a. *Sideritis scardica* Griseb. ssp. *scardica* (Figs. 5A, 6, 11D, 12E, 13).

Type collection: see under the species.

This subspecies is very common on Mt Vermion. There is no specific locality in Dimonie's collection from Mt Kajmakčalan which is a schistose mountain itself. It may come from a limestone branching of the above mountain.

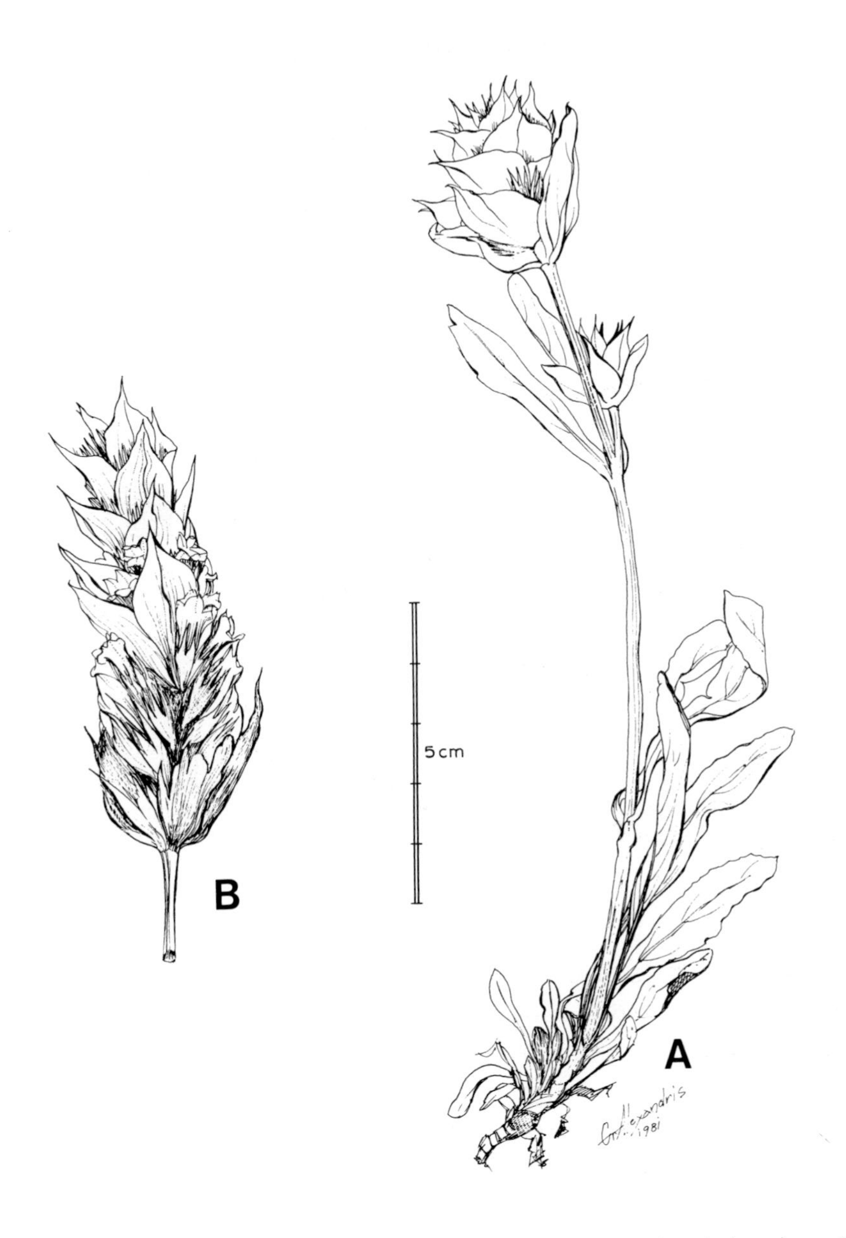

FIGURE 5. A: Habit of *Sideritis scardica* ssp. *scardica* (Lectotype). - B: Inflorescence of *S. scardica* ssp. *longibracteata*. Indumentum is not indicated.

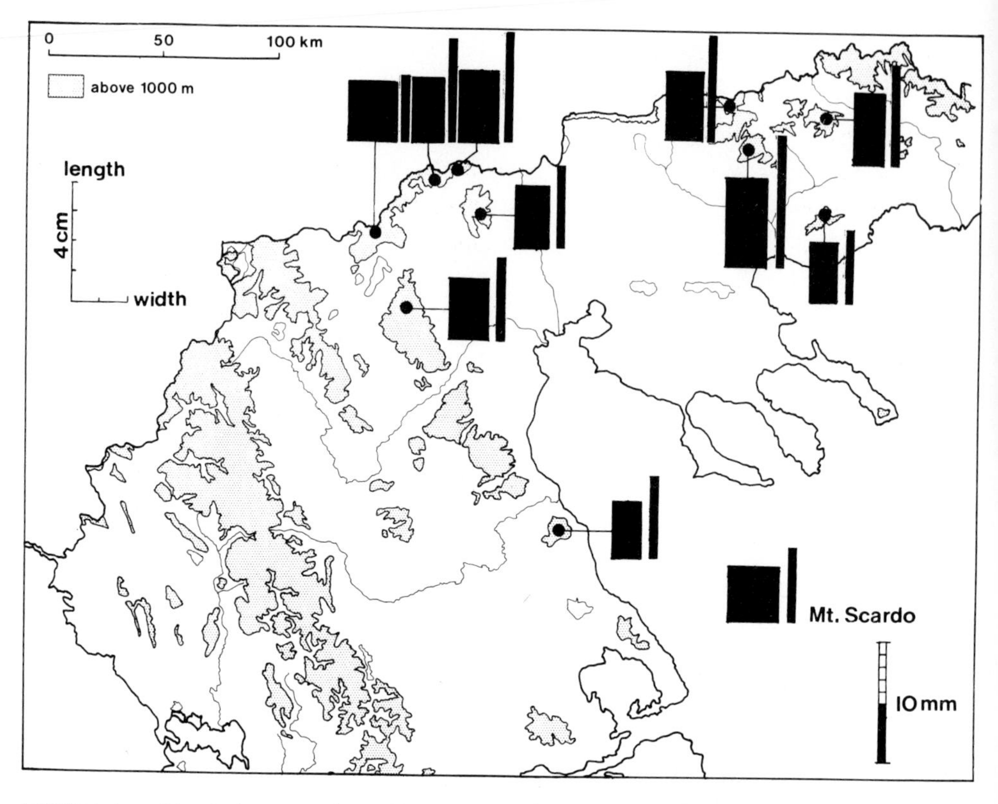

FIGURE 6. *Sideritis scardica*. Histograms (to the left) showing the average mean values of the bract length (vertical axis) and width (horizontal axis) from different parts of the distribution area. The length (to the right) of acumen. The subdivision of the species is to some extent based on these characters. All measurements are taken from material collected in nature.

Collections: Mt Scardo. Grisebach ? (K, Göt). - Mt Kajamakčalan (Voras). Dimonie 6/09 (W). - Mt Vermion. Rechinger 8871, 1.7.1936 (W); Gustavsson & Franzén 8547, 19.7.1979 (C), 8666, 20.7.1979 (LD).

3b. *Sideritis scardica* Griseb. ssp. *longibracteata* Papanicolaou & Kokkini ssp. nov. (Figs. 5B, 6, 12F, 13).

Orig. coll.: Nom. & Ep. Seron: Mt Menikion, SW part, ca 5.5 km NE of the village of Inousa (above the hamlet of Chionochorion), 1350-1600 m. Rocky subalpine meadows. Limestone. Strid & Papanicolaou 15774, 23.7.1979 (Holotype in C, isotype in ATH).

Inflorescentia 5-8 cm longa. Folia floralia viridescentia, anguste ovata ad oblonga, adpresse villosa, 19-40 X 9-14 mm magna.

This plant is somewhat taller than ssp. *scardica*, with longer inflorescence, and greenish, appressed hairy, narrowly ovate to oblong bracts. It is widespread in Greece occurring on Mt Ossa, Mt Tzena, Mt Pinovon, Mt Paikon, and in mountains of NE Greece (Fig. 13). The plants from NE Greece are very distinct from the type collection of ssp. *scardica*.

Collections: Mt Menikion. Strid & Papanicolaou 15774, 23.7.1979 (C); Rechinger 11036, 15.7.1936 (G, K, LD, W). - Mt Pangaion. Papanicolaou 102, 27.7.1977 (C, H.Th.), Papanicolaou 107, 27.7.1977 (C, H.Th.); Stainton 7784, 25.6.1959 (K, W); Strid 734, 9.7.1970 (C); Nordestam 6560, 14.7.1973 (C); Greuter 15328, 28.7.1977 (G); Rechinger 10206, 26.6.1936 (W), 10351, 26.6.1936 (W). - Mt Orvilos. Strid & Georgiadou 13274, 25.7.1977 (C). - Mt Falakron. Stamatiadou 10110, 24.7.1970 (W); Strid 1032, 19.7.1970 (C); Pinatzi 17489, 18.7.1961 (K); Rechinger 10658, 7.7.1936 (W). - Mt Tzena. Strid & Papanicolaou 16694, 19.8.1979 (C, LD); Greuter 14070, 30.7.1976 (G). - Mt Pinovo. Greuter 14640 (G); Strid & Papanicolaou 16542, 16.8.1979 (C). - Mt Paikon. Greuter 13988, 29.7.1976 (G). - Mt Ossa. J.S. Andersen 10157, 28.7.1975 (C); Grehenohivoff, 27.7.1936 (K).

4. *Sideritis raeseri* Boiss. & Heldr., Diagn. Pl. Orient. Ser. 2, 3: 30 (1854).

Type: In rupestribus reg. media et superioris Parnassi, alt. 4000'-6500'. Heldreich 490, 1856 (G lectotype selected here, syntypes in K, U, W).

Syn.: *Sideritis syriaca* Fraas, Syn. Pl. Fl. Clas. p. 175 (1845), non L.

Description: Perennial, grey-lanate, 10-50 cm tall, herb. Base of stem somewhat woody, producing several ascending, unbranched flowering shoots. Basal leaves moderate thick, densely lanate, crenate or entire, spathulate or oblong-spathulate, obtuse or subobtuse, attenuate into a petiole, 12-60 X 5-8 mm (including petiole). Upper stem leaves subsessile, rounded to attenuate at base, lanceolate or linear-lanceolate, acute, 10-50 X 5-7 mm. Inflorescence 3-21 cm long. Verticillasters 3-15, lax or very distant. Bracts covered by capitate glandular and small, patent eglandular hairs outside, shorter, equal or longer than flowers, reniform or ovate-orbicular or ovate in outline, 5-25 X 8-22 mm. Calyx dense-hairy all over with hairs 2-4 mm long, narrowly campanulate, 6-13 mm long. Corolla yellow without two stripes on the two lobes of the upper lip, 8-15 mm long.

Flowering period: From beginning of June to mid-August.

Distribution: *S. raeseri* occurs in mountains of Sterea Hellas, Pindhos range, with a single locality in NC Greece on Mt Olympos (Fig. 13) extending into Albania as well as Jugoslavia.

Habitat: It is a plant of alpine and subalpine zone, growing in stony places in meadows. Ssp. *attica* (see below) is found in both rocky places and rock crevices.

Variation: The species has been divided in three subspecies (see below), viz. ssp. *raeseri*, ssp. *attica* and ssp. *florida*. Plants most closely matching the type collection are found from Mt Parnassos northwards. A rather distinct geographical race, ssp. *attica*, occurs in the southernmost part of its distribution (Fig. 13). Plants from there are generally small, densely lanate, resembling *S. syriaca* in some extent. Plants from Olympos, ssp. *florida*, are taller than those of ssp. *attica* with inflorescence having lax verticillasters and larger bracts (Figs. 9, 12K). Ssp. *raeseri* has very distant verticillasters, bracts ovate-orbicular, something between those of ssp. *attica* and ssp. *florida* in size and shape (Figs. 9, 12I).

3.1.3. Key to subspecies

1. Verticillaster distant, bracts reniform or ovate-orbicular...........2
- Verticillaster lax to compact, bracts ovate in outline... ssp. *florida*
2. Plants 20-50 cm tall, bracts ovate-orbicular, equal or longer than flowers, 11-20 X 11-19 mm (length: width = 1-1.2)........... ssp. *raeseri*
- Plants 10-20(30) cm tall, densely lanate, bracts reniform, mucronate, shorter than flowers, 6-12 X 9-12 mm (length : width = 0.7-0.9)..........
.... ssp. *attica*

4a. *Sideritis raeseri* Boiss. & Heldr. ssp. *raeseri* (Figs. 7C, 9, 10, 11G, 12I, 13).

Synonymy and type: see under the species.

This subspecies occurs from Mt Parnassos northwards (Fig. 13). It is characterized by large, petiolate, oblong-spathulate, 12-60 X 5-8 mm, basal leaves, distant verticillasters with bracts equal or slightly longer than flowers, and calyx length something between ssp. *attica* and ssp. *florida* (8-10 mm).

Collections: 120 sheeths were examined.

4b. *Sideritis raeseri* Boiss. & Heldr. ssp. *attica* (Heldr.) Hal.) Papanicolaou & Kokkini comb. et stat. nov. (Figs. 8B, 11H, 12J, 13).

Basionym: *Sideritis attica* Heldr., Delt. Physiogr. Themat. Syll. Parn.

p. 46 (1900).

Syn.: *S. raeseri* Boiss. & Heldr. var. *attica* (Heldr.) Hal., MBL, 11, 179.

Orig. coll.: In saxosis regionis abietinae mont Parnethis, prope cacumen. Heldreich, Aug. 1901 (Herb Thessaloniki lectotype selected here, W).

This is a rather distinct geographical race which is endemic to the southern part of the species distribution. It is a densely lanate plant, shorter than ssp. *raeseri*, characterized by spathulate basal leaves, and bracts shorter than flowers. The calyx is hairy all over, shorter than that of both ssp. *raeseri* and ssp. *florida*. It occurs on Mts Pateras, Kitheron and Parnis.

Collections: Mt Parnitha. Rechinger 18160, 17.7.1956 (W); Stojanoff et Jordanoff, 16.7.1937 (K); Heldreich 1677, Aug. 1901 (W). - Mt Pateras. Pichler ? (W). - Mt Kitheron. Zaganiaris ? (Herb. Thessaloniki).

4c. *Sideritis raeseri* Boiss. & Heldr. ssp. *florida* (Boiss. & Heldr.) Papanicolaou & Kokkini comb. et stat. nov. (Figs. 8A, 11I, 12K, 13).

Basionym: *Sideritis florida* Boiss. & Heldr., Diagn. Plant. Orient. Nov. Ser. 2, 3: 31 (1854).

Syn.: *S. scardica* Griseb., Halácsy in Consp. Flor. Graec. 2: 498 (1902) pro parte; Boissier in Flor. Orient. 4: 710 (1879) pro parte; Heywood in Flor. Eur. 3: 142 (1972); Strid (21) in Wild Flowers of Mt Olympus.

Orig. coll.: In regione sylvatica Olympi Thessali prope caenobium Santi Dionysii. Heldreich 2517, 23.7.1851 (W lectotype selected here, syntypes in K).

This taxon is endemic to Mt Olympos in NC Greece. It occurs from an altitude of ca 900 up to 2800 m. Plants from lower altitudes are larger in all respects than those of alpine regions. Generally this subspecies is an intermediate between ssp. *raeseri* and *S. scardica* ssp. *scardica*. The inflorescence is longer than that of ssp. *scardica*; verticillasters lax, and bracts glandular-puberulent outside. As the feature of glandular capitate hairs outside bracts is characteristic for *S. raeseri* the authors have treated it as a subspecies of the latter. Boissier (3) & Heldreich have treated it as a separate species.

Plants from Mt Pilion (CE Greece), known as *S. scardica* Griseb. var. *pelia* Hal., have a resemblance in all respects to those of Mt Olympos. They have no glandular capitate hairs on bracts, however. The Pilion material may merit a subspecific rank of *S. scardica*, but more material is needed.

Collections: Mt Olympos. Sintenis 1881, 12.9.1889 (K, LD, W); Stojanoff & Jordanoff, 26.7.1937 (K); Orphanides 539, 28.7.1889 (K, W); Grebenchinoff, 4.8.1933 (K); Dibowski 1.7.1928 (K, G, W); Sintenis & Bornmüller 1429, 30.7.1891 (G, K, LD, W); Adamović, 20.7.1905 (G, K); Regel 2.8.1931 (G); Strid 1543, 29.7.1970 (C), 1276, 23.7.1970 (C); Strid & Hansen 9050, 7.8.1975 (C), 9408, 17.8.1975 (C); Hayek 2.6.1926 (W); Handel-Mazzeti 19.7.1927 (W); Heldreich 2517, 20.7.1851 (K, W), 2.9.1853 (K), 25.7.1882 (K).

5. *Sideritis syriaca* L., Sp. Pl. 574 (1753) (Figs. 7A, 11E, 12G, 13).

Note: Only the picture of the Linnaean type specimen of *S. syriaca* has been seen by authors. The habitat is just "in Crete".

Syn.: *S. cretica* Boiss., in Flor. Orient. 4: 708 (1879), non L.

Perennial, densely white-lanate all over, herb, 30-50 cm tall. Basal leaves thick, densely lanate, entire, oblong-spathulate or spathulate, obtuse, rather abruptly contracted into a petiole, 25-40 X 7-10 mm (including petiole). Upper stem leaves linear-lanceolate or linear, entire, subsessile, mucronate, 10-50 X 3-10 mm. Inflorescence 7-25 cm long. Verticillasters very distant. Bracts densely white-lanate outside, broadly ovate in outline, shorter than flowers, 7-14 X 7-10 mm. Calyx dense-hairy all over with hairs 4-6 mm long, campanulate, 6-9 mm long. Corolla pale yellow, without stripes on the two lobes of the upper lip, just exceeding calyx-teeth.

Flowering period: From mid-June to the end of July.

Distribution: *S. syriaca* is endemic to the mountains of the island of Crete (Fig. 13).

Habitat: Dry stony places above 1000 m.

Variation and affinities: *S. syriaca* is a fairly distinct species, characterized by its tall, robust habit, with long inflorescence and distant verticillastes. The whole plant is white-lanate. There is a superficial resemblance to *S. euboea*, which, however, is a shorter plant, with shorter, compact inflorescence, and larger, obovate basal leaves. *S. raeseri* ssp. *attica* comes, also, into consideration. It is, however, characterized by shorter stems, shorter inflorescence, and smaller bracts.

Collections: 75 sheeths were examined.

6. *Sideritis euboea* Heldr., Delt. Physiogr. Themat. Syll. Parn. 4 (1900) (Figs. 7B, 11F, 12H, 13).

FIGURE 7. Habit of: A: *Sideritis syriaca*. - B: *S. euboea* (Lectotype). - C: *S. raeseri* ssp. *raeseri* (Lectotype).

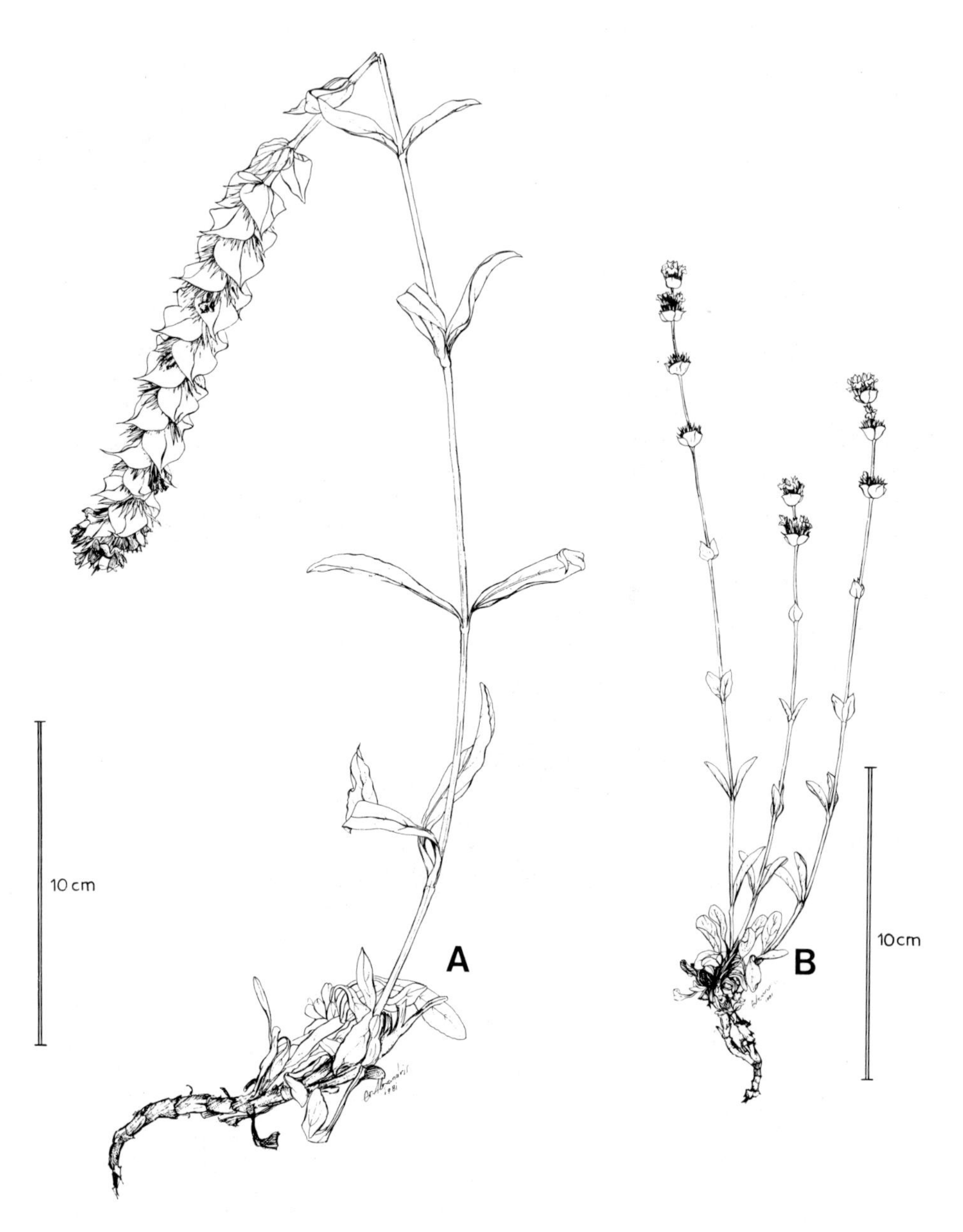

FIGURE 8. Habit of A: *Sideritis raeseri* ssp. *florida*. - B: *Sideritis raeseri* ssp. *attica*. Indumentum is not indicated.

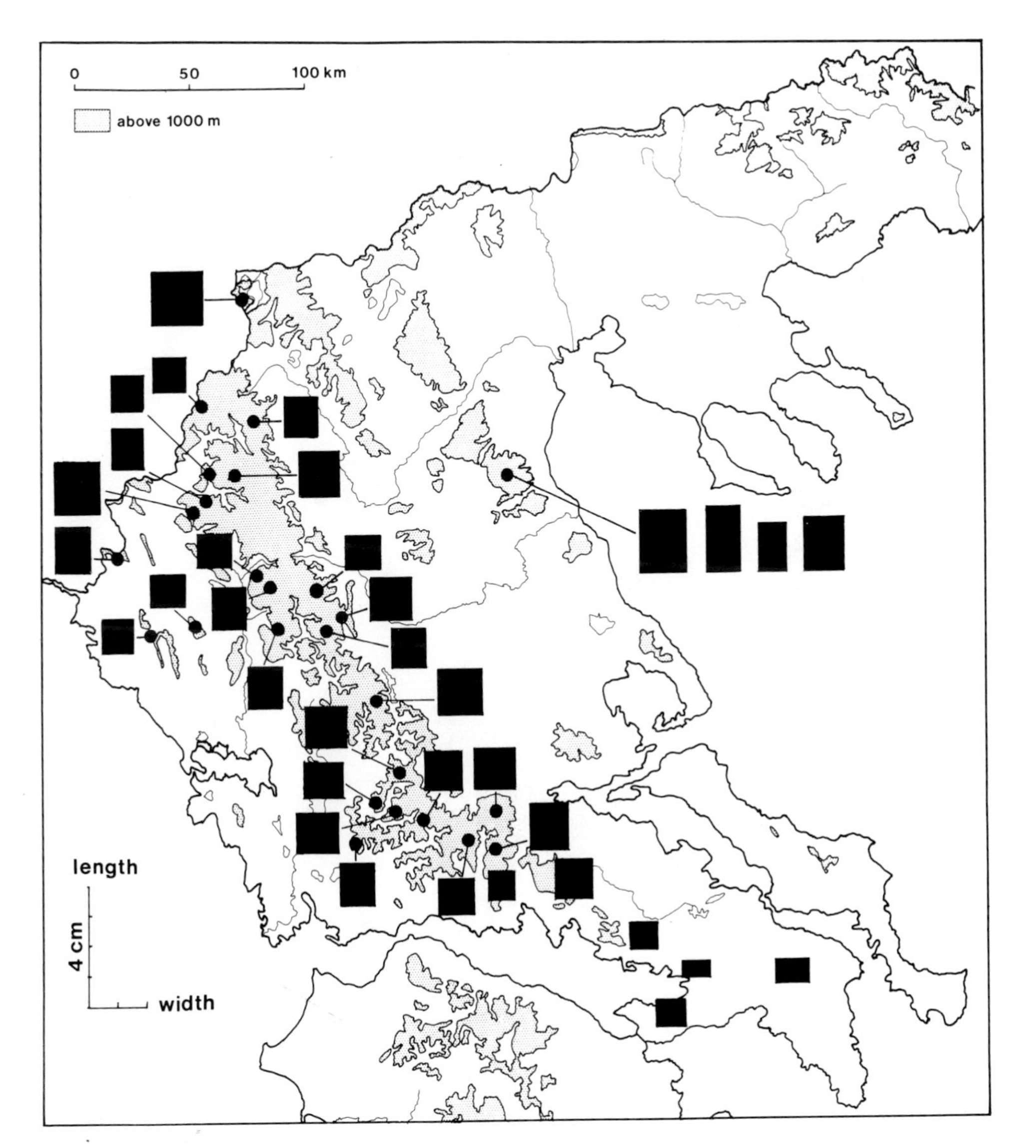

FIGURE 9. *Sideritis raeseri*. Histograms showing the average mean values of the bract length (vertical axis) and width (horizontal axis) from its distribution area in Greece. The subdivision of the species is to some extent based on this character. All measurements are taken from material collected in nature.

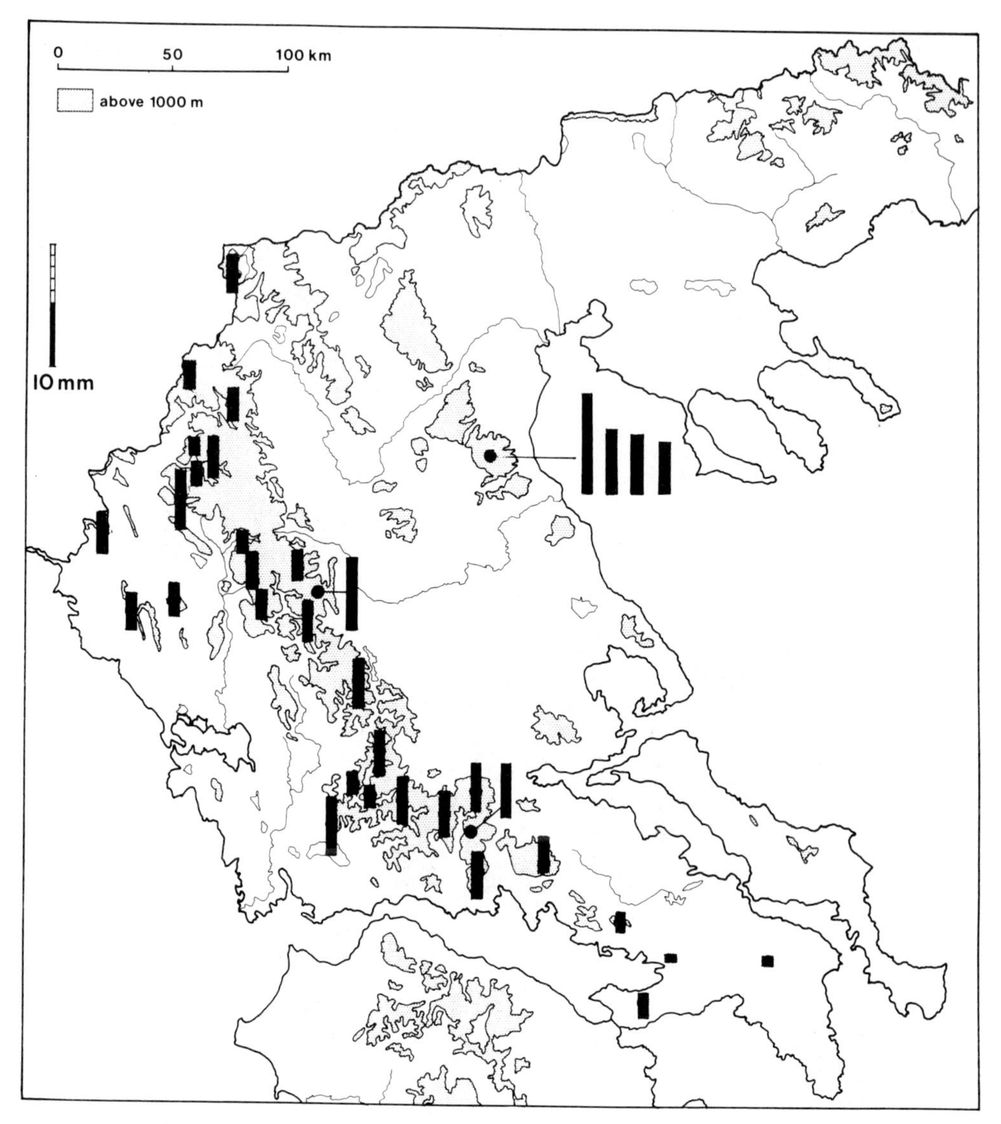

FIGURE 10. *Sideritis raeseri*. Length of acumen. The subdivision of the species is to some extent based on this character. All measurements are taken from material collected in nature.

Type: In regione superiori montis Delphi Euboea (Dirphys) inter lapides alt. 4500'-5100'. Heldreich 793 (K lectotype selected here, syntype in G).

Syn.: *S. cretica* Sibth. & Sm., in Prodr. Fl. Graec. 1: 400 (1806); *S. syriaca* L. var. *condesata* Boiss. & Heldr., Diagn. Pl. Orient. Nov. Ser. 2, 3: 31 (1859).

Perennial, densely white-lanate all over, herb, 15-35 cm tall. Basal leaves thick, densely lanate, entire or crenate, ovate-oblong or oblong, obtuse, rather abruptly contracted into a petiole, 20-50 X 10-15 mm. Upper stem leaves subsessile to sessile, oblong or obovate, shorter than internodes, 8-15 X 4-10 mm. Inflorescence 3-9 cm long. Verticillasters compact (the first two from below usually distant). Bracts usually lanate outside, broadly ovate, 7-12 X 7-10 mm. Calyx dense-hairy all over, campanulate, 7-9 mm long. Corolla yellow without stripes on the two lobes of the upper lip, just exceeding calyx-teeth.

Flowering period: From mid-June to the end of July.

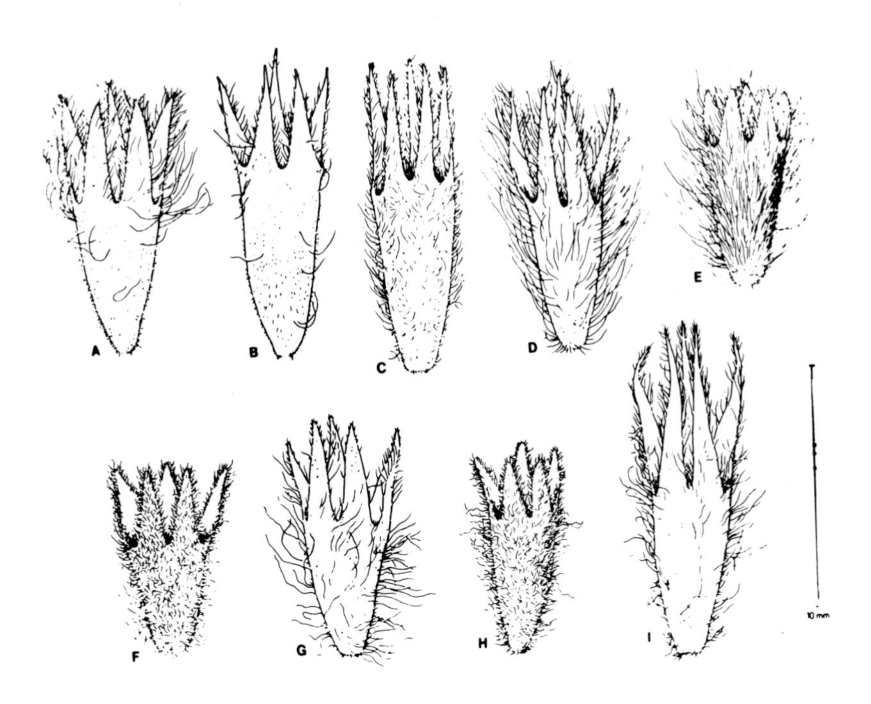

FIGURE 11. Calyx of: A: *Sideritis athoa*. - B: *S. perfoliata*. - C: *S. clandestina*. - D: *S. scardica*. - E: *S. syriaca*. - F: *S. euboea*. - G: *S. raeseri* ssp. *raeseri*. - H: *S. raeseri* ssp. *attica*. - I: *S. raeseri* ssp. *florida*.

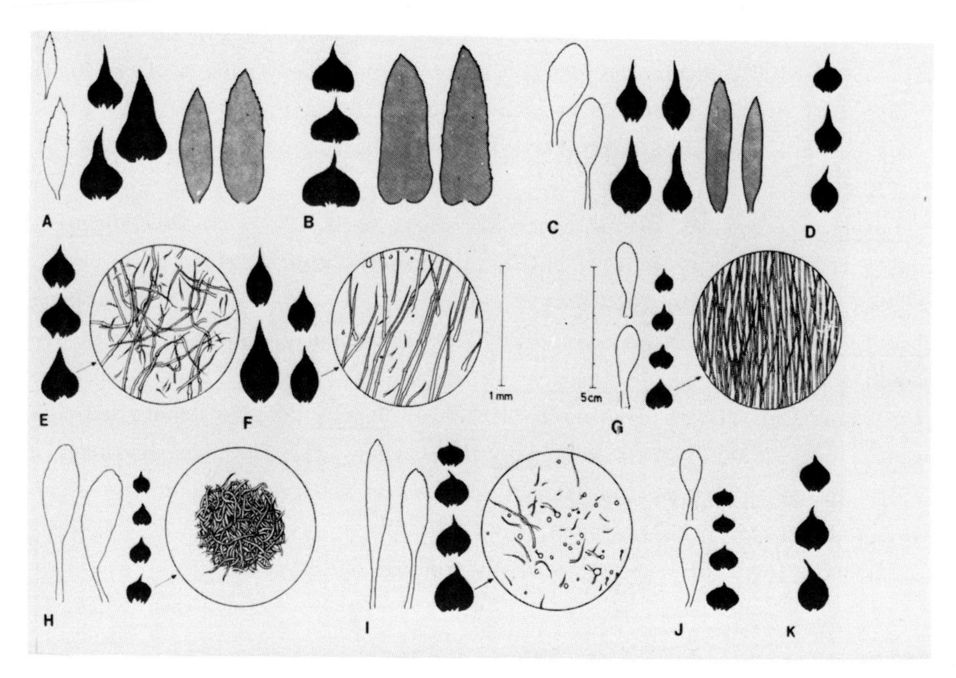

FIGURE 12. Basal leaves (outlined), stem leaves (grey)and bracts (black) of: A: *Sideritis athoa*. - B: *S. perfoliata*. - C: *S. clandestina* ssp. *clandestina*. - D: *S. clandestina* ssp. *cyllenea*. - E: *S. scardica* ssp. *scardica*. - F: *S. scardica* ssp. *longibracteata*. - G: *S. syriaca*. - H: *S. euboea*. - I: *S. raeseri* ssp. *raeseri*. - J: *S. raeseri* ssp. *attica*. - K: *S. raeseri* ssp. *florida*.

Distribution: *S. euboea* is endemic to the mountains (Mt Dirphys and Mt Xirovouni) of the island of Euboea (Fig. 13).

Habitat: Dry stony places above 600 m.

Variation and affinities: *S. euboea* is a relative of *S. syriaca* (see above). It is a rather distinct species, characterized by a short, compact inflorescence, and bracts densely lanate outside. There are, however, some specimens from Mt Dirphys having bracts glandular-puberulant outside, a character of *S. raeseri*.

Collections: 45 sheeths were examined.

⟶

FIGURE 13. The distribution of the *Sideritis* taxa section Empedoclia in Greece. ⊙ : *S. syriaca*, ◣ : *S. clandestina* ssp. *clandestina*, ■ : *S. clandestina* ssp. *cyllenea*, ▲ : *S. raeseri* ssp. *attica*, ⊙ : *S. raeseri* ssp. *raeseri*, ● : *S. raeseri* ssp. *florida*, ❋ : *S. scardica* ssp. *scardica*, ★ : *S. scardica* ssp. *longibracteata*, △ : *S. euboea*.

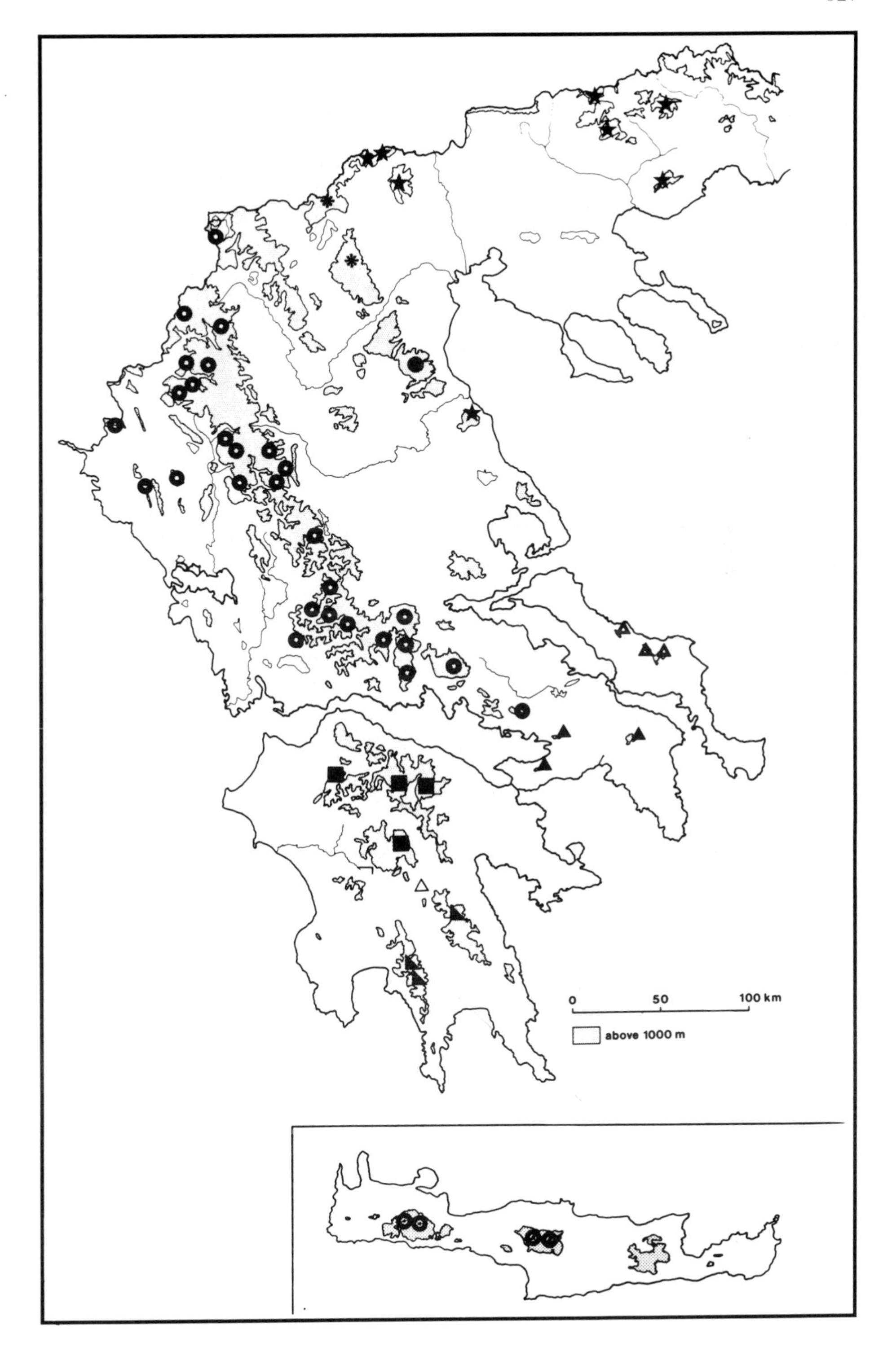
0
50
100 km
above 1000 m

REFERENCES

1. Aldén B. 1976. *Bot. Notiser* 129:297.
2. Boissier E. 1846. Diagnoses Plantarum Orientalium Novarum, Ser. 1, 7 (58). Lipsiae.
3. Boissier E. 1859. Diagnoses Plantarum Orientalium Novarum, Ser. 2, 3 (31-32). Lipsiae.
4. Boissier E. 1879. Flora Orientalis 4:714.
5. Chaubard LA. 1838. Nouvelle flore du Péloponnèse et des Cyclades. Strasbourg.
6. Chaubard LA, Bory S. 1832. Expéd. scient. de Morée, 170, t. 20
7. Contandriopoulos J. 1978. *Pl. Syst. Evol.* 129:277.
8. Dahlgren R. 1968 . *Bot. Notiser* 121:505.
9. Grisebach A. 1844. Spicilegium Florae Rumelicae et Bithynicae 2:144.
10. Halácsy E. von 1901-1904. Conspectus Florae Graecae 1-3. Lipsiae.
11. Hayek AV. 1929. *Feddes Repert. Sp. Nov. Beih.* 30(2).
12. Heywood VH. 1972. *Sideritis* L.In: Flora Europaea, Tutin TG, Heywood VH, Burges NA, Moore DM, Valentine DH, Walters SM (eds.) Vol. III p. 142. London, New York, Cambridge Univ. Press.
13. Hooker W. 1933. Hooker's Icones plantarum. Kew Herbarium, ser. 5, vol. 32, t. 3157.
14. Lanjouw J, Stafleu FA. 1964. Index Herbariorum, 1. The herbaria of the world. Regn. Veg. 31.
15. Linnaeus C. 1753. Species Plantarum Ed. 1. Holmiae.
16. Maire R, Petitmengin M. 1908. *Bull. Soc. Sc. Nancy*: 149, 360.
17. Rechinger KH. 1943. Flora Aegea. *Denkschr. Akad. Wiss. Wien. Math.-Nat. Kl.* 105:507.
18. Regel C. de. 1942. *Candollea* 9:104.
19. Stearn WT. 1973. Botanical Latin. ed. 2. London and Edinburgh.
20. Strid A. 1972. *Op. Bot.* 32:1.
21. Strid A. 1981. Wild Flowers of Mount Olympus. Kifissia-Athens.
22. Turrill WB. 1935. *Kew Bull.* 4:55.
23. Weimarck H. 1941. *Lunds Univ. Årsskr. N. F. Avd.* 2, 37:5.
24. Zaganiaris D. 1938-40. *Epist. Epet. Schol. Phys. Math. Panepist. Thessaloniki* 5:151.

CHAPTER 3

Chemotaxonomy

MORPHOLOGICAL, CYTOLOGICAL AND CHEMICAL INVESTIGATIONS OF *MENTHA SPICATA* L. IN GREECE.

S. KOKKINI, V.P. PAPAGEORGIOU

1. INTRODUCTION

Two years ago, we started to study the species of the genus *Mentha*, section Spicatae, growing in Greece, from a biosystematic point of view. This paper deals with the first results of our investigations on *Mentha spicata* L., which is the most variable species of the section Spicatae.

The large variation of the morphological characters of *M. spicata* was the reason for which many systematic botanists have attributed to it a number of names and subdivided it to many subspecies and varieties. Harley (1) has argued recently that the species *M. microphylla* C. Koch is not a distinct species but a subspecies of *M. spicata*, *M. spicata* ssp. *tomentosa*. The morphological differences between the two subspecies of *M. spicata* are mainly in the characters of the leaves (indumentum, dimensions, shape and margins).

According to the same botanist these two subspecies have the same parents, *M. longifolia* (L.) Hud. and *M. suaveolens* Ehr., but different origins. The origins of ssp. *spicata* are obscure, but may well have been in cultivation somewhere in the Near East at a remote period of antiquity. *M. spicata* ssp. *tomentosa* probably arose within the Aegean region and is more or less restricted to it.

2. MORPHOLOGICAL VARIATION

In Greece, glabrous or sparsely hairy forms of ssp. *spicata* are very often cultivated in gardens as culinary plants. These plants in some cases escape in nature but rarely persist long in any place. We have collected them mainly in the proximity of inhabited areas and only in few cases away from them.

M. spicata ssp. *tomentosa* is widely distributed in all Greek area and in all altitudes. It is always hairy and its leaves have numerous branched hairs in the lower surface. Plants of ssp. *tomentosa* in Greece have a very variable appearance. From field experiments we believe that it is due either

Margaris N, Koedam A, and Vokou D (eds.): Aromatic Plants: Basic and Applied Aspects

ISBN 90-247-2720-0.
Printed in the Netherlands.

to the segregation of its parental characters or to the hybridization with ssp. *spicata* which occurs in some cases when the two subspecies come into contact.

It is thus evident that in several Greek populations no distinction between these two subspecies can be drawn from only morphological characters. At present, we consider that it is more appropriate to consider the E Mediterranean hairy plants as belonging to the ssp. *tomentosa*, and the cultivated and naturalised glabrous or sparsely hairy plants to the ssp. *spicata*. The chromosome numbers of individuals from 20 localities of Greece were found to be 2n=48 for both subspecies.

3. CHEMICAL VARIATION

The study of the chemical variation of the species *M. spicata* is based on the volatile oil compounds. We have chosen these compounds since adequate knowledge on the main pathways, responsible for the formation of the main constituents is already available both in genetic and biochemical terms (2).

We have collected plants of the two subspecies from NC Greece area and having in mind the results of biogenetic studies, we have tried to identify the mono- and sesquiterpenoid constituents of the essential oils using a Gas chromatographic-Mass spectrometric computerized system.

3.1 *Mentha spicata* ssp. *spicata*

All the individuals of this subspecies studied have carvone and dihydrocarvone as main constituents. In the investigated populations, variation of the main constituents is only quantitative. This is true both for the cultivated individuals and for the native ones, in the few cases where they were found.

The native individuals of ssp. *spicata* that were studied are growing in damp areas, in low altitudes (up to 450 m) and they are glabrous or in some cases sparsely hairy.

The gas chromatogram of the essential oil of ssp. *spicata* from a population near the Lake of Kastoria is shown in Fig. 1. The results obtained from a GC-MS analysis can be seen in Table 1.

Table 1. Chemical composition of essential oil of *Mentha spicata* ssp. *spicata*.

No of peak	R_t (min)	Components	%
1	3.77	α-Pinene	0.62
2	4.74	Camphene	0.05
3	5.91	β-Pinene	1.30
4	7.93	Myrcene	2.66
5	9.83	Limonene	0.40
6	10.31	1.8-Cineole	6.88
7	12.01	Ocimene	0.16
8	13.11	γ-Terpinene	0.06
9	14.55	*p*-Cymene	0.02
10	15.64	Unidentified	traces
11	16.49	Terpinolene	0.01
12	19.35	3-Octyl acetate	0.01
13	22.82	Aldehyde	0.15
14	25.23	Amyl valerate Dehydro-*p*-cymene	0.02
15	26.55	Hexyl valerate	0.21
16	28.34	Hexenyl valerate	0.01
17	29.00	α-Copaene	0.51
18	30.94	β-Bourbonene	4.06
19	32.61	Unidentified	0.32
20	34.09	Unidentified	0.71
21	35.31	Dihydrocarvone	17.67
22	36.07	Caryophyllene	1.10
23	37.13	Unidentified	0.80
24	37.75	Unidentified	0.68
25	38.39	Unidentified	0.45
26	39.21	Dihydrocarvyl acetate Terpinyl acetate	10.14
27	42.11	Carvone and Naphthalene (traces)	39.75

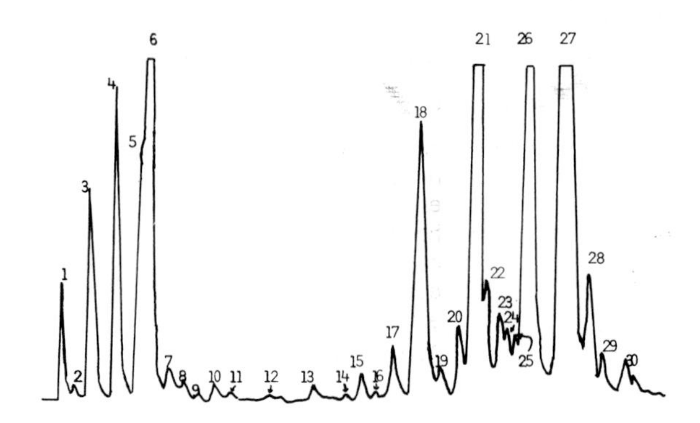

FIGURE 1. Gas chromatogram of the essential oil of *Mentha spicata* ssp. *spicata*.

3.2. *Mentha spicata* ssp. *tomentosa*

Three chemotypes of this subspecies have been distinguished up to now:
a. Chemotype 1: The individuals of this chemotype were found in two places of the studied area growing in low altitudes (up to 400 m) and in subhumid places.

These individuals have large leaves, lanceolate-oblong, deep green, hairy beneath, with undulate to serrate margins and apex subacute. The inflorescence is compact or interrupted below, especially in fruit.

The gas chromatogram of the essential oil of the above mentioned chemotype, from a population near the city of Kavalla, is shown in Fig. 2. The results obtained from a GC-MS analysis can be seen in Table 2. The main constituent is piperitenone oxide.

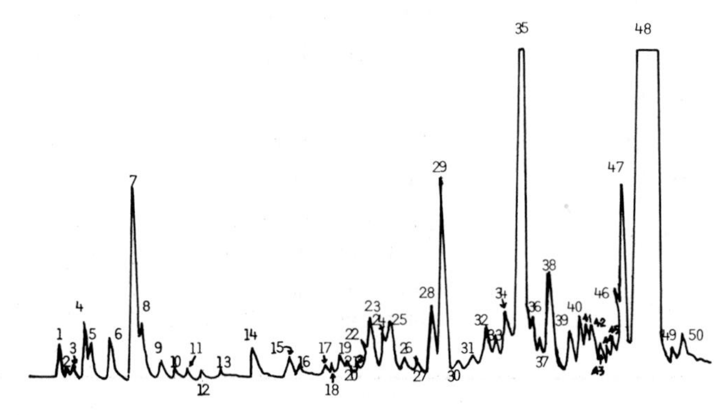

FIGURE 2. Gas chromatogram of the essential oil of *Mentha spicata* ssp. *tomentosa* (Chemotype 1).

Table 2. Chemical composition of essential oil of *Mentha spicata* ssp. *tomentosa* (Chemotype 1).

No of peak	R_t (min)	Components	%
1	3.79	α-Pinene	0.12
2	4.23	Unidentified	0.01
3	4.75	Camphene	0.02
4	5.91	β-Pinene	0.32
5	6.28	Unidentified	0.22
6	7.90	Myrcene	0.35
7	9.74	Limonene	1.68
8	10.27	1.8-Cineole	0.41
9	11.92	γ-Terpinene	0.13
10	13.06	*p*-Cymene	0.03
11	14.13	Terpinolene	0.002
12	15.07	Unidentified	0.01
13	16.61	Unidentified	0.03
14	19.24	3-Octyl acetate	0.26
15	22.20	3-Octanol	0.14
16	22.89	Nonanal	0.08
17	25.21	Dehydro-*p*-cymene	0.03
18	25.72	Unidentified	0.001
19	26.23	Menthone	0.13
20	26.72	Unidentified	0.05
21	27.71	Octyl acetate	0.05
22	28.03	Isomenthone Hexenyl isovalerate	0.01
23	28.55	α-Copaene	0.74
24	29.67	Unidentified	0.25
25	30.13	β-Bourbonene	0.64
26	31.37	Unidentified	0.17
27	32.25	Unidentified	0.12
28	33.41	Bornyl acetate	0.66
29	34.23	Carhydranol	2.00
30	35.51	Pulegone	0.15
31	36.71	Unidentified	0.28
32	37.76	Unidentified	0.56
33	38.58	Unidentified	0.42
34	39.35	Unidentified	0.80
35	40.41	Germacrene D	6.88
36	41.39	Unidentified	0.44
37	41.85	Unidentified	0.21
38	42.82	δ-Cadinene	1.21
39	44.47	Unidentified	0.36
40	45.25	Unidentified	0.55
41	45.75	Unidentified	0.36
42	46.12	Unidentified	0.34
43	46.39	Unidentified	0.22
44	46.86	Unidentified	0.19
45	47.61	Unidentified	0.20
46	48.21	Unidentified	0.30
47	48.53	Piperitenone	1.51
48	50.31	Piperitenone oxide	67.98

b. Chemotype 2: This is the most common chemotype in the area studied. Individuals of this chemotype were found in all altitudes from 30 up to 1400 m in humid places.

These individuals have small leaves, ovate to elliptic, dull green, densely hairy beneath with undulate to serrate margins and apex subacute. The inflorescence is lax, compact or intermediate.

The gas chromatogram of the essential oil of this chemotype from a population of Mt Vernon (1300 m), is shown in Fig. 3. The main constituent of this oil is piperitone oxide.

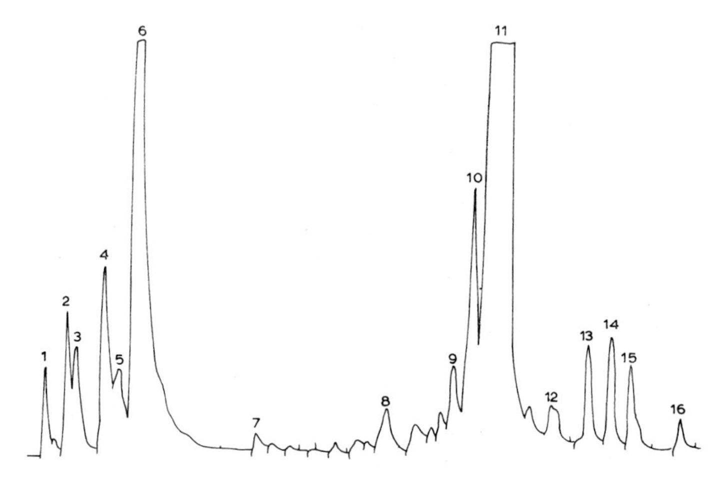

FIGURE 3. Gas chromatogram of the essential oil of *Mentha spicata* ssp. *tomentosa* (Chemotype 2).

c. Chemotype 3: Individuals of this chemotype have been found in two places of the studied area near villages and in waste places.

The plants of this chemotype have small leaves, ovate to obovate, dark green, more or less hairy beneath, with serrate to flat margins and apex acute. The inflorescence is more or less lax.

The gas chromatogram of the essential oil of this chemotype from a population near the village Chortiatis (alt. 400 m) is shown in Fig. 4. The results obtained from a GC-MS analysis can be seen in Table 3. The main constituent is linalool.

This chemotype in a novel one of the species *Mentha spicata*.

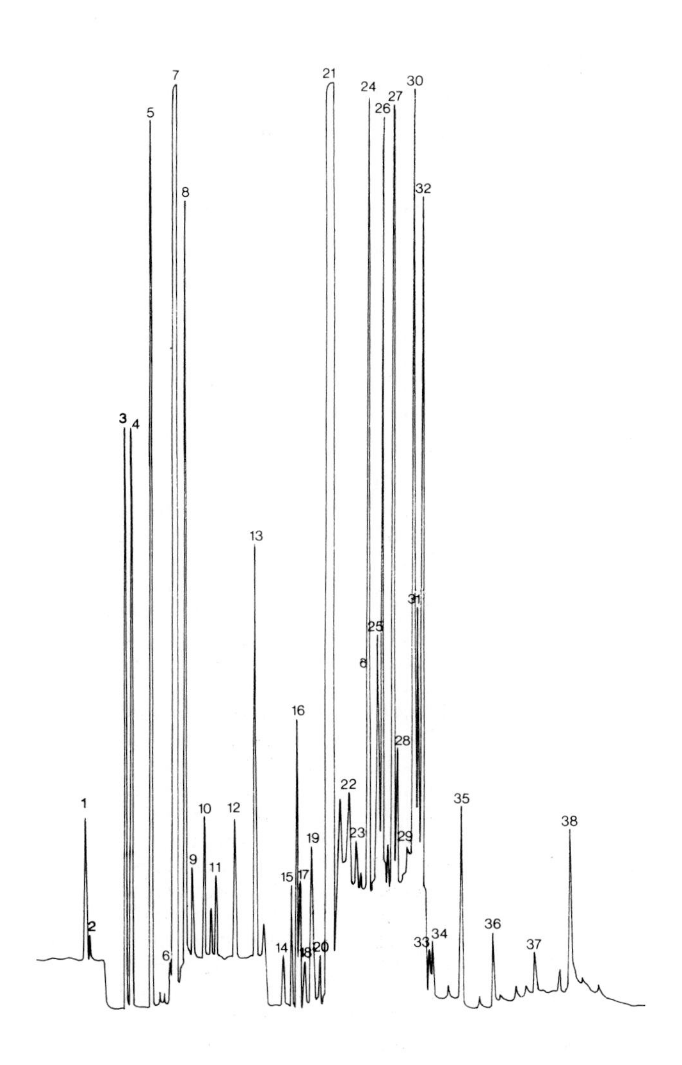

FIGURE 4. Gas chromatogram of the essential oil of *Mentha spicata* ssp. *tomentosa* (Chemotype 3).

Table 3. Chemical composition of essential oil of *Mentha spicata* ssp. *tomentosa* (Chemotype 3).

No of peak	R_t (min)	Components
1	9.10	α-Pinene
2	9.40	Toluene
3	11.20	β-Pinene
4	11.50	Sabinene
5	12.60	Myrcene
6	13.70	Limonene
7	14.00	1.8-Cineole
8	14.60	*cis*-Ocimene
9	15.00	γ-Terpinene
10	15.70	*p*-Cymene
11	16.30	Amyl valerate
12	17.40	3-Octyl acetate
13	18.50	3-Octanol
14	20.10	Linalool oxide (I)
15	20.50	Hexyl valerate
16	20.80	Epoxy linalool-2
17	21.00	Unidentified
18	21.20	3-Hexenyl valerate Octyl acetate
19	21.70	Isomenthone
20	22.10	Unidentified
21	22.40	Linalool
22	23.80	Unidentified
23	24.10	1-Terpinen-4-ol
24	24.70	β-Caryophyllene
25	25.30	*trans*-Pinocarveol
26	25.50	α-Terpineol
27	26.10	β-Terpineol
28	26.30	Unidentified
29	26.90	Unidentified
30	27.10	Germacrene-D
31	27.40	Naphthalene
32	27.60	δ-Cadinene
33	28.10	isom. Cadinene (tr.)
34	28.20	Nerol
35	29.80	Geraniol
36	31.40	Carveol
37	33.70	Farnesyl acetate
38	35.50	Thymol or carvacrol

CONCLUSIONS

1. From the morphological investigations of the Greek populations and field experiments conducted up to now, we are in agreement with Harley that *Mentha microphylla* C. Koch should be considered as a subspecies of *M. spicata*.

2. It is found that besides the morphological differences there is a chemical distinction between the two subspecies *spicata* and *tomentosa*. Thus, in the ssp. *spicata* the main constituents of the essential oil are carvone and dihydrocarvone, while in the ssp. *tomentosa* piperitenone oxide, piperitone oxide and linalool (chemotypes 1, 2 and 3, respectively).

3. The three chemotypes that we found in ssp. *tomentosa* are related to the environmental conditions and the morphological appearance of the plants (Table 4).

Table 4. Morphological and ecological differences between three chemotypes of ssp. *tomentosa* in NC Greece.

	Chemotype 1	Chemotype 2	Chemotype 3
Leaves:	40-56X11-15mm, lanceolate-oblong, deep green, hairy beneath, with undulate to serrate margins, apex subacute	20-30X10-12mm, ovate to elliptic, dull green, dense hairy beneath, with undulate to serrate margins, apex subacute	26-30X10-12mm, ovate to obovate, dark green, some hairy beneath, with serrate to flat margins, apex acute
Inflorescence:	compact or interrupted below especially in fruit	lax, compact or intermediate	$\pm$lax
Biotope:	subdamp places, from 30 up to 450 m	subwet places, from 30 up to 1400 m	waste places, from 100 up to 400 m

All the above characters may suggest further subdivision of the species *Mentha spicata*.

4. The fact that we found individuals of the species *spicata* having as main constituent of their essential oil piperitenone oxide should be related to the presence of *M. suaveolens* in these altitudes. *M. suaveolens* is the one parent of *M. spicata* and has been also found to contain piperitenone oxide as its main constituent (3). The individuals with piperitone oxide as their main constituent are growing both in low and high altitudes. *M. longifolia*, the other parent of *M. spicata* growing in high altitudes,

contains also piperitone oxide as its main constituent (3).

On the contrary, carvone and linalool were not found up to now in any of the two parents present in the investigated area.

REFERENCES

1. Harley RM. 1982, in press. *Notes R. Bot. Gard. Edinburgh*.
2. Hefendehl FW, Murray MJ. 1976. *Lloydia* 39:39.
3. Kokkini S, Papageorgiou VP. 1981. Unpublished data.

DISTRIBUTION OF FLAVONOIDS AS CHEMOTAXONOMIC MARKERS IN THE GENUS *ORIGANUM* L. AND RELATED GENERA IN LABIATAE

S.Z. HUSAIN, V.H. HEYWOOD, K.R. MARKHAM

1. INTRODUCTION

Origanum L. is a taxonomically complex genus containing approximately 26-30 species. Its geographical distribution extends from North Africa and Western Europe to Central Asia. One species *Origanum vulgare* has a very wide distribution from the Azores to Taiwan, whereas most species are endemic to a single island or mountain. The centre of diversity is in the Mediterranean basin where over 70% of the species are found. They usually occur in mountainous regions ranging from 500 to 1500 metres in altitude in rocky places with calcareous soils. The species of *Origanum* like so many other Mediterranean labiates, are distinctly shrubby or at least woody at the base, and rich in volatile oils. These oils are found in glands which are distributed mainly in the leaves. The genus has been variously treated in the past, some authors recognizing a number of distinct genera e.g. (2), others recognizing only sections e.g. (1): both treatments are still in current use (3, 11).

2. PROCEDURE

2.1. Material and Methods

In the course of a phytochemical survey of *Origanum* and related genera, dry leaf-material from herbarium specimens and from fresh samples of *Origanum* species grown from seeds in the Plant Science Laboratories glasshouse complex was used.

30 species of *Origanum* were examined for flavones, flavanols, flavone C-glycosides and cinnamic acids in leaf tissue. Most of the flavonoid constituents were completely characterized. Representative species of related genera were also screened for their flavonoid constituents.

2.1.1. Two-dimensional paper chromatography of direct leaf extracts. One gram of fresh leaves, or 100 mg of dried material was ground and extracted 2 x with MeOH-H_2O (4 : 1). 2D-PCs (Whatman No 1) were spotted with the extract from ca 50 mg of plant material (dry wt) and run in n-BuOH-

Margaris N, Koedam A, and Vokou D (eds.): Aromatic Plants: Basic and Applied Aspects
 ISBN 90-247-2720-0. Printed in the Netherlands.

-HOAc-H_2O (4:1:5, BAW) and 15% HOAc. UV-absorbing spots were circled then fumed with ammonia vapour, any colour changes recorded and Rf values were calculated (BAW and 15% HOAc). Tables 1, 2, 3 show the distribution of spots 1-19 in the *Origanum* and related genera species with their Rf values in five different solvents and tentative identification of flavone glycosides.

2.1.2. Identification of glycosides by acid-hydrolysis. For the aglycone moiety of the glycosides and the identity of sugars, a small portion of the purified sample was treated with 2N HCl at $100^{\circ}C$ for 1.5 hours (for glucuronides) or the sample was treated with 2N HCl at $100^{\circ}C$ for 30 minutes (where no glucuronides were suspected). After acid hydrolysis, resultant mixture was extracted with ethyl acetate (1 : 1), the upper EtOAc layer containing the (EtOAc) aglycone moiety, and the lower aqueous phase containing sugars, were concentrated separately. The concentrate of the upper layer was diluted with 1-2 drops of 95% EtOH (ethanol), spotted onto Whatman No 1 chromatography paper, run one-dimensionally in the solvents; n-BuOH-HOAc-H_2O (4:1:5, BAW), conc. HCl-HOAc-H_2O (3:3:10, Forstal), phenol-H_2O (4 : 1), with the aglycone markers.

The lower aqueous layer which contained sugars was dried on a rotatory evaporator to remove HCl, then dissolved in deionized water (1-2 drops) and chromatographed one dimensionally on Whatman No 1 paper in solvents phenol-H_2O (4 : 1), n-BuOH-benzene-pyridine-H_2O (5:1:3:3, B B P W). Reference sugars were glucose, galactose, xylose, arabinose, rhamnose and glucuronic acid. The developed chromatograms were dipped in aniline hydrogen phthalate reagent (6), dried, and heated ($110^{\circ}C$) for 5-10 minutes. Sugar spots appeared brown or orange.

The identity of flavonoid glycosides was established by a combination of per-methylation, mass and UV spectroscopy and co-chromatography with authentic markers.

2.1.3. Identification by UV spectroscopy. Glycoside samples were dissolved in 100% MeOH and their UV absorption spectra measured between 225 and 450 nm (Pye Unicam recording spectrophotometer). Spectra were recorded before and after the addition of small amounts of diagnostic inorganic reagents (aqueous sodium hydroxide (2N), solid anhydrous sodium acetate and boric acid, or 5% aluminum chloride in EtOH). The wavelength maximum shifts, according to the location of various functional groups on the aromatic (A/B) rings (6, 7). (Fig. 1 shows a typical representation of UV

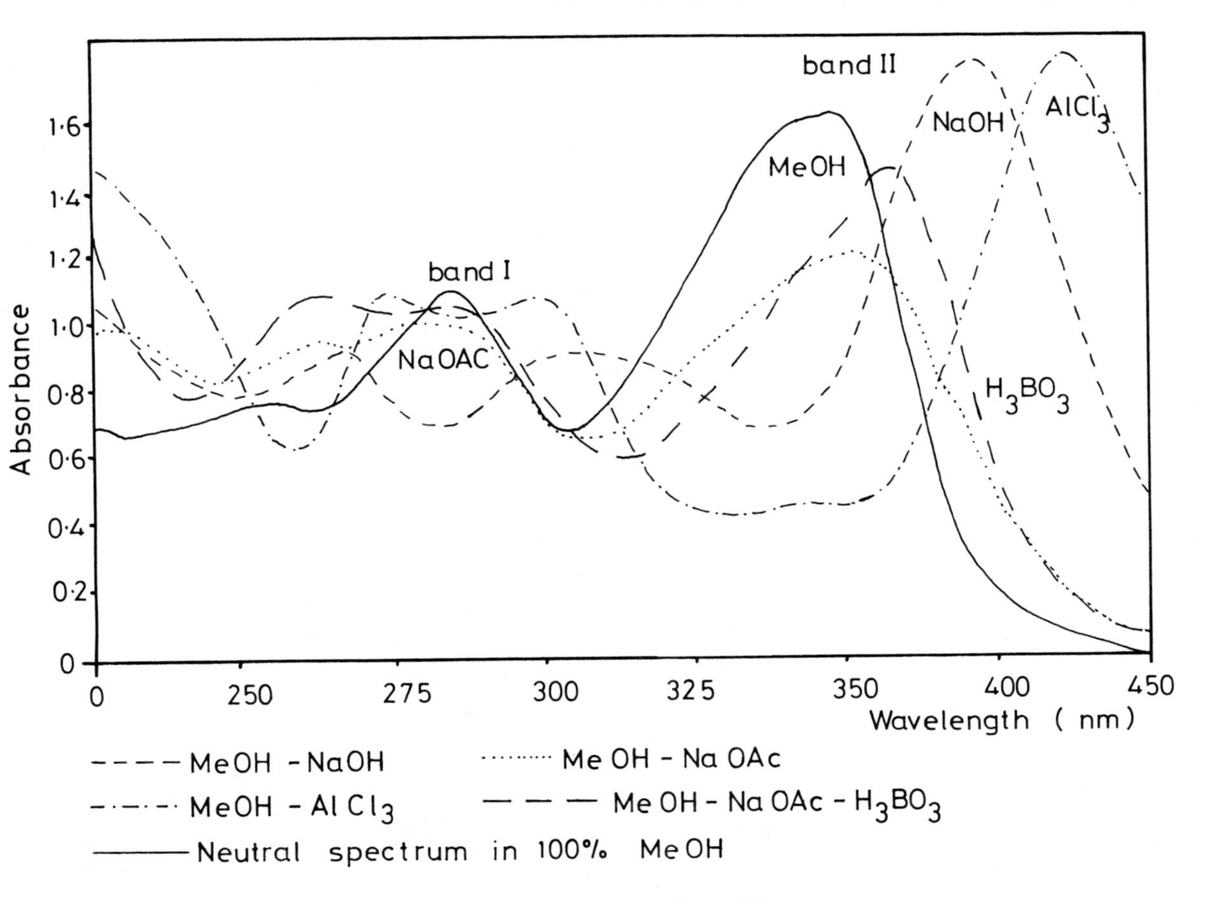

FIGURE 1. UV Spectra of 6-hydroxyluteolin 7-glucoside.

spectra of a 6-hydroxyluteolin 7-glycoside of *Origanum* species).

3. RESULTS

In the course of a chemotaxonomic survey of *Origanum* and other allied genera, in which at least 75% of the species of each genus was examined, 20 phenolic compounds were observed and recorded on two dimensional (2D) chromatograms. The most common leaf-constituents were flavone O-glycosides. No flavonols were detected in this survey and the biosynthetically distinct flavone C-diglycoside, apigenin 6,8-C-diglucoside (vicenin-2) was characterized in only 17% of the sample and only in certain taxonomic groups (Table 4). Thus (V-2) is present in all species of section Majorana of the genus *Origanum* and in both sections Pseudothymbra and Thymus of the genus *Thymus*, and in three *Hyssopus* species, but is apparently absent in species of sections Amaracus and Origanum of the genus *Origanum* and the species of *Lycopus, Rosmarinus, Calamintha, Melissa* and *Micromeria* examined which contain only the flavone O-glycoside.

4. DISCUSSION

The present investigations confirm that luteolin is of common occurrence in genera of Labiatae (8). Luteolin 7-glycoside was observed in 93%, the 7-diglucuronide in 90% and 7,4'-diglucoside in 63% of the species surveyed. Luteolin 5-glucoside however, was present in only 13% of the sample. Two apigenin glycosides were characterized, apigenin 7-glucoside (found in 40% of the species) and apigenin 7-glucoronide (present in 13% of the sample) chrysoeriol (luteolin 3'-methyl ether) occurred as the 7-glucuronide in 23% and 6-hydroxyluteolin was recorded in 50% of the species as 7-glucoside. Scutellarein 6,4'-dimethyl ether occurred free, i.e. not glycosylated in 15 species of *Origanum*.

From the data presented in Table 1, it is clear that the presence or absence of derivatives of luteolin, apigenin, chrysoeriol and scutellarein allows the species of *Origanum* to be conveniently divided into six groups.

Apart from vicenin-2, five *Origanum* species fall in group one: *O. dubium, O. majorana, O. onites, O. syriacum* and *O. microphyllum* containing 6-hydroxyluteolin 7-O-glucoside, luteolin 7-diglucuronide, luteolin 7-O-glucoside, apigenin 7-O-glucuronide, 4'-O-methyl-scutellarein and 6,4'-di-O-methyl-scutellarein. In addition, chromatograms of *O. majorana* leaf extracts showed two unidentified spots (15 and 17), plus a third attributable to luteolin 7-O-rutinoside. Spot 15 was also associated with *O. syriacum* and

O. onites. Within the Majorana group only *O. microphyllum* contained either luteolin 7,3'-O-glucuronide or glucoside and luteolin 7,4'-O-diglucoside and lacked the unidentified spots 15 and 17.

In all other species of *Origanum*, vicenin-2 is absent. The second group with the following species, *O. vulgare, O. heracleoticum, O. virens* and *O. viride*, although lacking vicenin-2, generally seems to show a common pattern with luteolin 7-diglucuronide, luteolin 7,3'-O-glucuronide (or glucoside), luteolin 7,4'-O-diglucoside and luteolin 7-O-glucoside being present in these species when examined. 6-Hydroxyluteolin 7-O-glucoside was also present additionally in *O. virens* and *O. viride*. Luteolin 7-O--rutinoside, 4'-O-methylscutellarein, 6-4'-O-dimethyl ester of scutellarein and an unidentified spot (15) were also present in this group. Apigenin glycosides were not detected.

The third group which consists of only *O. compactum, O. laevigatum* and *O. ehrenbergii*, seems uniform in that whilst all species lack scutellarein methyl ethers and apigenin glycosides, they contain luteolin 7-rutinoside, luteolin 7,3'-O-glucuronide (or glucoside), luteolin 7,4'-diglucoside and luteolin 7-O-glucoside, and unidentified spot 16.

The fourth group similarly contains 3 species, *O. isthmicum, O. ramonense,* and *O. dayi* but, in contrast to group 3, exhibits a very complex flavonoid profile. Thus, all three species have 6-hydroxyluteolin 7-O-glucoside, scutellarein-6,4'-dimethyl ether, luteolin 7-diglucuronide, apigenin 7-O-glucuronide, luteolin 7-O-glucoside, apigenin 7-O-glucoside and chrysoeriol 7-glucuronide. In addition, *O. isthmicum* also has luteolin 7,4'-diglucoside and luteolin 5-O-glucoside: these flavonoids are absent from *O. ramonense*, which contains unidentified spot 17 not found in either *O. dayi* or *O. isthmicum*.

The fifth group includes *O. amanum, O. cordifolium, O. ciliatum, O. dictamnus, O. libanoticum, O. micranthum, O. saccatum, O. sipyleum* and *O. tournefortii*. Most of these species contain luteolin 7,3'-O-glucuronide or glucoside, luteolin 7-O-diglucuronide, luteolin 7-O-glucoside, luteolin 7,4'-diglucoside. Although 4'-O-methylscutellarein is also common in many species of this group, it is noticeably absent from *O. amanum, O. cordifolium* and *O. libanoticum*. In *O. sipyleum*, apigenin 7-O-glucoside, in *O. dictamnus* apigenin 7-O-glucuronide and the 6,4'-dimethyl ether of scutellarein are also present with *O. ciliatum* and *O. saccatum* the only two species in this group containing luteolin 5-O-glucoside and *O. amanum* additionally

Table 1. Flavonoid glycoside spots detected in species of *Origanum*.

Taxon Spot No.*	1	2	3	4	5	6	7	8	9
Origanum dubium	+	+	-	-	-	+	-	-	+
O. majorana	+	+	-	-	-	+	+	-	+
O. microphyllum	+	+	(+)	+	-	+	-	-	+
O. onites	+	+	-	-	-	+	-	-	+
O. syriacum	+	+	-	-	-	+	-	-	+
O. heracleoticum	-	+	+	+	+	+	-	-	-
O. virens	(+)	+	+	+	+	+	-	(+)	-
O. viride	(+)	+	+	+	+	+	-	-	-
O. vulgare	-	+	(+)	+	-	+	(+)	-	-
O. compactum	-	-	+	+	+	-	+	-	-
O. ehrenbergii	-	-	+	-	+	-	+	-	-
O. laevigatum	-	+	+	+	+	+	+	-	-
O. dayi	+	+	-	-	-	+	+	-	+
O. isthmicum	+	+	-	(+)	-	+	-	+	+
O. ramonense	+	+	-	-	-	+	-	-	+
O. amanum	-	-	+	+	+	-	-	-	-
O. cordifolium	-	+	+	+	+	+	-	-	-
O. ciliatum	-	+	(+)	+	+	+	-	+	+
O. dictamnus	-	+	-	+	-	+	-	-	-
O. libanoticum	-	+	+	+	+	+	-	-	-
O. micranthum	-	(+)	-	-	-	(+)	-		-
O. saccatum	-	(+)	+	+	+	+	-	+	(+)
O. sipyleum	-	+	+	-	+	+	-	-	-
O. tournefortii	-	(+)	+	+	+	+	-	-	-
O. akhdarense	(+)	+	-	+	-	+	-	-	-
O. hypericifolium	-	+	+	+	+	+	-	-	-
O. leptocladum	(+)	+	+	-	+	+	-	-	-
O. rotundifolium	(+)	+	+	+	+	+	-	-	-
O. scabrum	(+)	+	+	+	+	+	-	+	-
O. vetteri	+	+	-	-	-	+	-	-	+

*see Table 2.

†Spot 20 is present in *O. amanum*.

10	11	12	13	14	15	16	17	18	19	Maximum number of spots present
-	+	+	+	-	-	-	-	-	-	7
-	+	+	+	-	+	-	+	-	-	10
-	+	(+)	(+)	-	-	-	-	-	-	9
-	+	+	+	-	+	-	-	-	-	8
(+)	+	+	+	-	+	-	-	-	-	9
-	-	+	(+)	-	(+)	-	-	-	-	8
-	-	-	-	-	(+)	-	-	-	-	8
-	-	-	-	-	-	-	-	-	-	6
-	-	-	-	-	-	-	-	-	-	5
-	-	-	-	-	-	(+)	-	-	(+)	6
-	-	-	-	-	-	(+)	-	-	+	5
-	-	-	-	-	-	(+)	-	-	(+)	8
+	-	+	(+)	+	-	-	-	(+)	-	10
+	-	+	-	-	-	-	-	(+)	-	9
+	-	+	-	-	-	-	+	(+)	(+)	9
-	-	-	-	-	-	-	-	-	-	4†
-	-	+	-	-	-	-	-	-	-	6
+	-	(+)	(+)	+	-	-	-	-	-	11
-	-	+	(+)	-	-	-	-	(+)	-	7
-	-	-	-	-	-	-	-	-	-	5
-	-	-	(+)		-	-	-	+	-	4
-	-	+	+	+	-	-	-	(+)	-	11
+	-	-	+	-	-	-	-	-	-	6
-	-	-	+	-	-	-	-	-	-	6
-	-	+	+	-	-	-	-	-	-	6
+	-	-	(+)	-	-	-	-	(+)	-	8
-	-	-	(+)	-	-	-	-	-	-	6
-	-	-	+	-	-	-	-	-	-	7
-	-	(+)	+	-	-	-	-	-	-	9
-	-	-	+	-	-	-	-	-	-	5

Table 2. The characteristics of flavone glycosides found in species of *Origanum* and related genera.

Spot No.	Colour in UV	Colour in UV + NH_3	Compound identified
1	D	D	6-hydroxyluteolin 7-O-glucoside
2	D	Bty	luteolin 7-O-diglucuronide
3	D	flLGn	luteolin 7,3'-O-diglucuronide or glucoside*
4	D	y	luteolin 7,4'-O-diglucoside
5	D	flLGn	luteolin or apigenin glucuronides *
6	D	Bty	luteolin 7-O-glucoside
7	D	y	luteolin 7-O-rutinoside
8	W/B	fly	luteolin 5-O-glucoside
9	D	dy	apigenin 7-O-glucuronide
10	D	dy	apigenin 7-O-glucoside
11	D	dOR	apigenin 6,8-di-C-glucoside (vicenin-2)†
12	D	D	scutellarein 6,4'-dimethyl ether
13	D	D	scutellarein 4'-methyl ether 7-O-glucoside
14	D	D	scutellarein 7-O-glucuronide
15	D	y	unidentified
16	D	dy	unidentified
17	D	Bty	unidentified
18	D	yGn	chrysoeriol 7-O-glucuronide
19	W	W	unidentified
20	D	dy	unidentified

* = tentative identification
† = also characterized by per-methylation and mass spectroscopy.
D = Dark; y = yellow; B = Blue; Gn = Green; L = Light;
Bt = Bright; OR = Orange; d = dull; W = White; fl = fluorescent.

Table 3. Rf values of spots 1-20 in five different solvents.

Spot No.	Rf values x 100				
	BAW	15% HOAc	H_2O	BEW	PhOH
1	18	6	1	28	44
2	23	11	18	-	20
3	20	23	63	-	4
4	17	32	60	-	-
5	15-20	25-40	40-90	-	5-20
6	32	15	1	46	59
7	38	34	7	-	62
8	31	6	1	36	45
9	48	24	20	44	43
10	53	22	3	54	80
11	30	54	-	-	57
12	78	11	-	-	88
13	54-80	5-8	2	-	60-79
14	35-50	6-12	30	-	30
15	8	44	-	-	-
16	10	65	-	-	-
17	6	24	-	-	-
18	44	17	32	-	50
19	30	12	-	-	-
20	50	70	-	-	-

BEW = n-BuOH-ethanol-water (4:1:2:2)

Table 4. Presence of apigenin 6,8-di-C-glucoside (vicenin-2) in *Origanum* species and selected species from related genera. Family Labiatae (subtribe Thyminae, tribe Saturejeae).

Genera	Section	Species	Location of investigated plants	Preser of vicen
Origanum L.	Majorana (Miller) T.Vogel	*O. majorana* L.	Spain,Italy	+
		O. microphyllum (Benth.) Boiss.	Crete	+
		O. onites L.	Greece,Spain	+
		O. syriacum L.	Libya,Jordan	+
		O. majoricum Camb.	Spain	+
		O. dubium Boiss.		+
Thymus L.*	Pseudothymbra Benth.	*T. longiflorus* Boiss.	Spain	+
		T. membranaceus Boiss.	Spain	+
		T. villosus L.	Spain,Portugal	+
		T. antoninae Rouy & Coincy	Spain	+
		T. mastigophorus Lacaita	Spain	+
		T. leucotribcus Halácsy	Crete	+
		T. cherlerioides Vis.	Albania,Bulgaria	+
		T. cephalotos L.	Portugal,Spain	+
	Thymus (section vulgares Velen., section zygis Willk.)	*T. vulgaris* L.	Spain	+
		T. capitellatus Hoffmann	Portugal	+
		T. zygis L.	Spain	+
		T. camphoratus Hoffmann	Portugal	+
		T. hyemalis Lange	Spain	+
		T. baeticus Boiss.ex Lacaita	Spain	+
		T. hirtus Willd.	Spain	+
Hyssopus L.		*H. officinalis* L.	Spain,France	+
		H. officinalis subsp. *montanus* (Jordan & Fourr.) Briq.	U.S.S.R.	+
		H. officinalis subsp. *canescens* (DC.) Briq.	Spain	+

* A few species of other sections of *Thymus* were examined. These do show the presence of V-2 but the sample studied was not large enough to include in this survey.

containing the unidentified component 20. Species in group five which lack 6-hydroxyluteolin 7-O-glucoside contained apigenin glycosides and chrysoeriol 7-glucuronide.

The sixth and final group consists of *O. akhdarense*, *O. hypericifolium*, *O. leptocladum*, *O. rotundifolium*, *O. scabrum* and *O. vetteri*. Here 6-hydroxyluteolin 7-O-glucoside, luteolin 7-O-diglucuronide, luteolin 7,3'-diglucuronide or -diglucoside are common to most species examined. In *O. hypericifolium*, apigenin 7-O-glucoside, is also present; *O. scabrum* contains additionally, luteolin 5-O-glucoside and the 6,4'-dimethyl ether of scutellarein. *O. hypericifolium* is the only species in group 6 lacking 6-hydroxyluteolin 7-o-glucoside and has only a trace of the 4'-methyl ether of scutellarein.

The presence of apigenin 7-O-glucuronide and particularly of apigenin 6, 8-di-C-glucoside (vicenin-2), is the first reported occurrence of flavone C-glycoside in the genus *Origanum* (9) and probably also in the Labiatae (7). Furthermore, its occurrence appears to be taxonomically significant and does not reflect any geographic pattern (Table 4) suggesting that it is probably independent of environment. Its presence in species of section Thymus and Pseudothymus of the genus *Thymus* and the genus *Hyssopus* indicates the close relationship of *Origanum* with these genera. Its presence only in *Majorana* species appears to provide a useful chemotaxonomic character at sectional level within the genus *Origanum*.

From the chemotaxonomic investigations carried out on the genus *Origanum* and on the representative species of related genera and from the data in Table 1, it can be seen that flavonoid distribution lends support to the currently accepted classifications of the species in sections Majorana and Origanum (3, 10), but does not support the separation off of *O. amanum* as a separate section (10) nor the division of the remaining species (section Amaracus) into four sections.

When further work from essential oil distribution and variation, palaeonological micro-characters (based on light and scanning electron microscopy), anatomy (cuticular studies) and other taxonomical characters of these species are taken into consideration it may be possible to subdivide these groups further.

REFERENCES

1. Boissier E. 1879. *Flora Orientalis* 4:546.
2. Briquet J. 1897. In: Die natürlichen Pflanzenfamilien, Engler A and Prantl K (eds.) 4:304. Leipzig, W Engelmann.

3. Fernandes R, Heywood VH. 1972. *Origanum* L. In: Flora Europaea, Tutin TG, Heywood VH, Burges NA, Moore DM, Valentine DH, Walters SM, Webb DA (eds.) vol III, p. 171. London, New York, Cambridge Univ. Press.
4. Jurd L. 1962. In: Chemistry of the Flavonoid Compounds, Geissman TA (ed.). Oxford, Pergamon Press.
5. Harborne JB. 1967. Comparative Biochemistry of the Flavonoids. London and New York, Academic Press.
6. Harborne JB. 1973. In: Phytochemical Methods. London, Chapman and Hall.
7. Harborne JB, Mabry TJ, Mabry H (eds.). 1975. The Flavonoids. London, Chapman and Hall.
8. Harborne JB, Williams CA. 1971. *Phytochemistry* 10:367.
9. Husain SZ, Markham KR. 1981. *Phytochemistry* 20:1171.
10. Ietswaart JH. 1980. A taxonomic revision of the genus *Origanum* (Labiatae). The Hague/Boston/London, Leiden University Press.
11. Rechinger KH fil. 1943. Flora Aegea. Wien.

NEW CHEMICAL MARKERS WITHIN *ARTEMISIA* (COMPOSITAE-ANTHEMIDAE)

H. GREGER

1. INTRODUCTION

The genus *Artemisia* comprises about 400 species mainly distributed in the northern hemisphere. It is the largest and most widely distributed of approximately 100 genera (18) in the tribe Anthemidae (Compositae). Within the genus there is a great variety of life forms from dwarf herbaceous representatives to tall shrubs occurring in a wide range of habitats between arctic alpine regions to dry deserts. Most of the *Artemisias* are characterized by the presence of strong aromatic odours mainly based on mono- and sesquiterpenes which are accumulated both in glandular hairs and schizogenous secretory canals.

Systematic arrangements have been based so far predominantly on general floral morphology leading to the traditional four sections Abrotanum Bess., Absinthium DC., Seriphidium Bess., and Dracunculus Bess. (6, 15) or the three subgenera *Artemisia* Less., *Seriphidium* (Bess.) Rouy, and *Dracunculus* (Bess.) Rydb. (23). However, more recent regional or sectional systematic treatments (7, 19, 21, 24) have shown that these infrageneric divisions are not entirely satisfactory.

Because of the worldwide use of different *Artemisia* species in herbal medicine many investigations have been carried out in order to determine their chemical composition. Most of these analyses were restricted to the aboveground parts and were mainly concerned with sesquiterpene lactones (for ref. see 20), monoterpene compositions (1, 8, 26), and flavonoid profiles (9, 16). However, based on the extensive investigations of Bohlmann and his collaborators within the Compositae it is evident that many typical secondary constituents of the family are often accumulated especially in the underground parts.

In the course of current comparative analyses on *Artemisia* root constituents, a preliminary survey of approximately 200 well documented species and provenances has shown that the distribution of different polyacetylenes (10, 11), coumarin-sesquiterpene ethers (14), and sesamin type lignans (12, 13) provides a valuable systematic criterion within the genus. The present

Margaris N, Koedam A, and Vokou D (eds.): Aromatic Plants: Basic and Applied Aspects
 ISBN 90-247-2720-0.
Printed in the Netherlands.

chapter demonstrates to what extent these differences may serve as chemical characters and contribute to a more natural arrangement within *Artemisia*.

2. RESULTS

2.1. Polyacetylenes

The polyacetylenes are products of the fatty acid synthesis and are accumulated in schizogenous secretory canals. Up to now, more than 600 different derivatives could be isolated mainly from the Compositae (2). Since the biogenetic pathways of many acetylenes have already been established they represent an excellent source of chemical characters for the whole family. Within the Anthemidae specific transformation processes lead to different groups of biogenetically closely related compounds which apparently reflect, to a large extent, the natural relationships of the tribe (9).

Most of these characteristic Anthemidae-polyynes could also be found in the genus *Artemisia*, where different biogenetic trends contribute to a species grouping which corresponds remarkably well with the traditional infrageneric classification. One of the most striking differences results from the vicarious occurrence of either dehydrofalcarinone I derivatives together with aromatic acetylenes II-IV (in the subgenus *Dracunculus* (11) and in the "Heterophyllae" group (7) of the subgenus *Artemisia*) or spiroketalenol-ethers V, VI, XI, pontica epoxide VII, and derivatives of artemisia ketone VIII and dehydromatricaria ester IX (in the remaining *Artemisia* species).

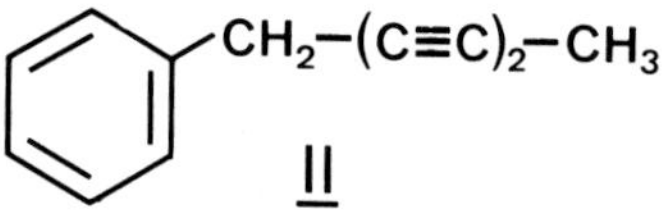

$-CH_2-(C\equiv C)_2-CH_3$

II

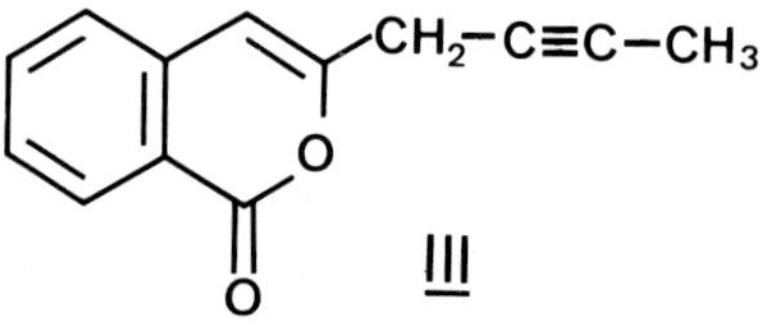

$-CH_2-C\equiv C-CH_3$

III

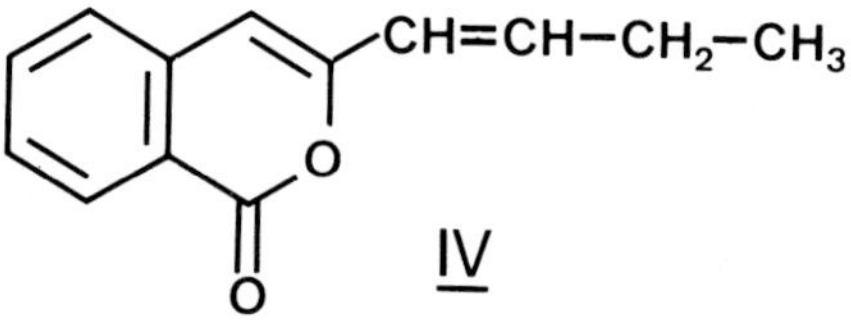

$-CH=CH-CH_2-CH_3$

IV

$CH_2=CH-C(=O)-(C\equiv C)_2-CH_2-CH=CH-(CH_2)_5-CH=CH_2$

I

Since all members of the subgenus *Dracunculus* are clearly separated from the other subgenera by their sterile central disc florets as well as by other morphological characters, this corresponding biosynthetic capacity with the "Heterophyllae" (with hermaphrodite disc florets) deserves special systematic consideration. Within the subgenus *Dracunculus* a distinction may be made on the basis of distinct accumulation tendencies either towards dehydrofalcarinone derivatives or aromatic acetylenes. The different provenances of *A. dracunculus* itself and some closely related species can be characterized by a preponderance of aromatic acetylenes (Table 1) (11).

Moreover, the dehydrofalcarinone pathway also dominates in the taxonomically rather isolated E-Asiatic *A. keiskeana* Miq. as well as in the arctic alpine *A. norvegica* Fries., both of which are recognized as members of the subgenus *Artemisia*. However, in spite of the hermaphrodite disc florets, the former species closely resembles the subgenus *Dracunculus*, especially with respect to its hemispheric involucre of flower heads. In the latter species there are so many morphological similarities with the "Heterophyllae" (7) that this chemical trend additionally supports an inclusion in this species group.

From the remaining *Artemisia* species the subgenus *Artemisia* (including the sections Artemisia, Abrotanum and Absinthium) produces the most structurally diversified acetylenes which are often accumulated in considerable amounts. The subgenus *Seriphidium* (excluding the N-American "Tridentatae"), by contrast, can be clearly separated by a strong decrease of acetylene accumulation. Besides the widespread derivatives of dehydromatricaria ester IX and artemisia ketone VIII different accumulation tendencies either towards spiroketalenol ethers V, VI, XI or pontica epoxide VII have been shown to be of special systematic significance. Both chemical trends are typical for the Anthemidae (9) and have not been found so far in other tribes of Compositae (2). Their biosynthesis can be considered to diverge at the stage of a triyne-enoic precursor X (Fig. 1).

Spiroketalenol ethers are widespread in members of the sections Artemisia and Absinthium and often represent the major biogenetic trend in, for example, the alpine *A. mutellina* group (*A. eriantha* Ten., *A. genipi* Weber, *A. mutellina* Vill. etc.) and the probably closely related N-American *A. scopulorum* A. Gray and *A. pattersonii* A. Gray (section Absinthium) Moreover, they may also dominate as thiophene VI-(i.e. in *A. verlotiorum*

Table 1. Distribution of root acetylenes in the subgenus *Dracunculus* and in the "Heterophyllae".
(● as major, + as minor biogenetic trend)

Artemisia species and provenances	Dehydro-falcarinone derivatives	Aromatic acetylenes		
		Capillen II	Capillarin III	Artemidin IV
subgenus *Dracunculus*				
crithmifolia L.	●			
capillaris Thunb.				
- N-Japan	●	+		
- S-Japan	●	●	+	
monosperma Del. (21)	●	●		
scoparia W. & K.	●	+		
commutata Bess.	●			
campestris L.	●			
bottnica A.N.Lundstr.	●			
borealis Pall.	●			
caudata Michx	●			
pycnocephala DC.	●			
desertorum Spreng.	●	●		
japonica Thunb.	●			
littoricola Kitam.	●			
dracunculus L.				
- USSR	+			●
- USSR	+			●
- W-Germany	+			●
- USSR	+	●		
- W-Germany	+	+	●	
- Finland	+	●		
- Canada	+	●	●	
- U.S.A.	+	●	●	
- N-India	+	●		
dracunculiformis Krasch.	+	●		
pamirica C.Winkl.	+	●		●
glauca Pall. ex Willd.	+	●		
"Heterophyllae"				
atrata Lam.	●			
laceniata Willd.	●			
armaniaca Lam.	●			+
pancicii Ronn.	●			
oelandica (Bess.) Krasch.	●			
latifolia Ledeb.	●	+		+
punctata Bess.	●			●

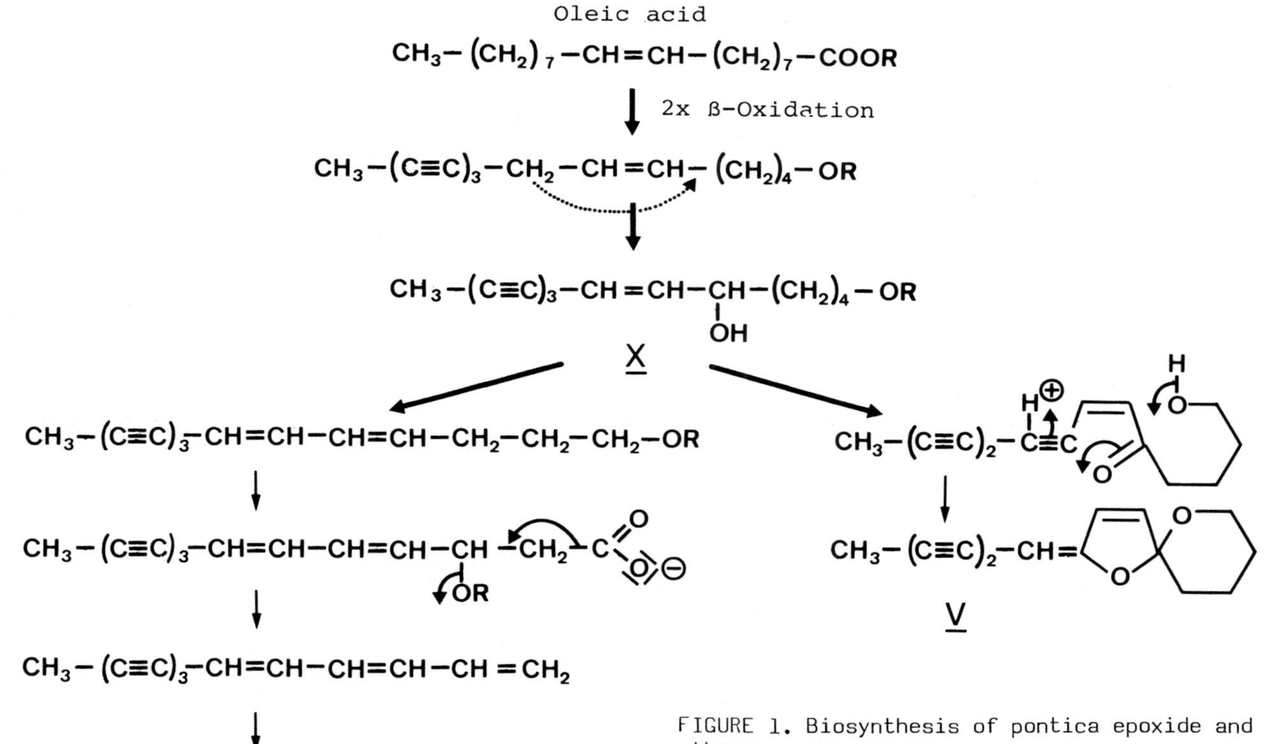

FIGURE 1. Biosynthesis of pontica epoxide and spiroketanol ethers.

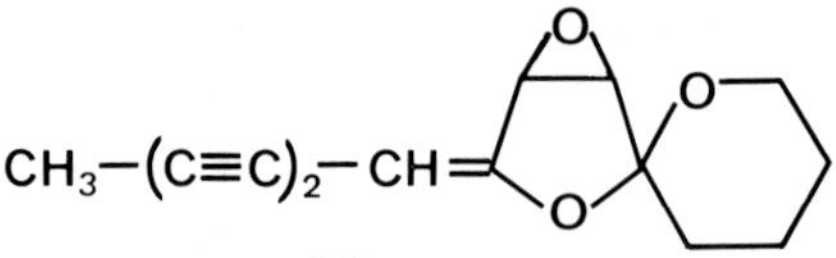

$CH_3-(C\equiv C)_2-CH=$ [spiroketal epoxide structure]

XI

[thienyl]$-CH=$ [spiroketalenol ether structure]

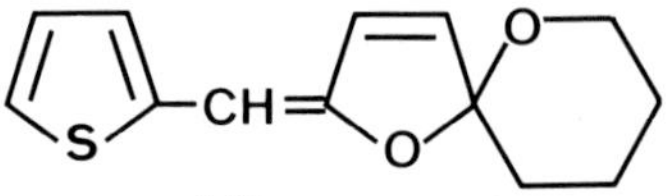

VI

$CH_3-(C\equiv C)_3-CH=CH-C(=O)-O-CH_3$

IX

$CH_3-(C\equiv C)_3-CH=CH-(CH_2)_2-C(=O)-CH_2-CH_3$

VIII

Lamotte, *A. incisa* Pamp., *A. ludoviciana* Nutt. (section Artemisia)) or epoxi-derivatives XI (i.e. in *A. douglasiana* Bess. (section Artemisia)).

In many members of the section Abrotanum the pontica epoxide pathway dominates (Fig. 1), apparently replacing the spiroketalenol ether accumulation. High concentrations of pontica epoxide have specifically been found in members of the Asiatic series Vestitae (*A. vestita* Wall. ex DC., *A. gmelinii* Web. ex Stechm. etc.) and in the probably closely related S-African *A. afra* Jacq. (5). Furthermore, this biogenetic trend also characterizes *A. pontica* L. and *A. persica* Boiss. both of which have been shown to possess nearly identical acetylenic equipment. Although the latter species had been considered to be part of the section Absinthium because of its hairy receptacles (23), Ehrendorfer (7) has suggested that it is closely related to the series Vestitae on the basis of similar leaf morphology. Its affinities with the section Abrotanum are supported by the present chemical findings.

All Eurasiatic members of the subgenus *Seriphidium* investigated so far are characterized by very small amounts of acetylenes without spiroketalenol ethers. In some species (i.e. in *A. lercheana* Web. ex Stechm., *A. caerulescens* L. ssp. *gallica* (Willd.) K.Pers.) the formation of pontica epoxide resembles that found in the section Abrotanum. Taxa within the N-American section Tridentatae (sensu Rydberg (24)), by constrast, can be clearly distinguished by their accumulation of spiroketalenol ethers (i.e. in *A. tridentata* Nutt. ssp. *tridentata, A. tripartita* Rydb. ssp. *tripartita*) and other acetylenic compounds. These data may be in agreement with a more recent proposal by McArthur et al. (22) in which the Tridentatae were treated as a separate subgenus.

Finally, in *A. annua* L. the distinct trend towards pontica epoxide accumulation, as well as various morphological characteristics, indicate a close relationship to the series Vestitae. On the other hand *A. biennis* Willd., *A. tournefortiana* Reichb., and *A. klotzschiana* Bess. which are considered to be closely related to *A. annua* (15, 23), can be separated by the large amounts of spiroketalenol ethers, which they possess.

2.2 Sesquiterpene-coumarin ethers

Until now, sesquiterpene-coumarin ethers have nearly exclusively been found in the gum resins of *Ferula* and some closely related umbelliferous genera (17). The coumarin moiety of all these derivatives has been shown to be uniformly umbelliferon (25). Considering this restricted distribution the detection of isofraxidin derived sesquiterpene ethers in *Artemisia pontica* L. and *A. abrotanum* L. (4) is of particular chemosystematic interest.

Extensive comparisons within *Artemisia* have shown that the formation of ethers with sesquiterpenes linked to isofraxidin represents a typical chemical trend in the section Abrotanum. Of the compounds identified so far the sesquiterpene moieties have proven to be either farnesyl XII or bicyclic drimenyl derivatives XIII (14). According to Bohlmann (4) drimartol A XIII has been isolated from *A. pontica* and *A. abrotanum* and additionally

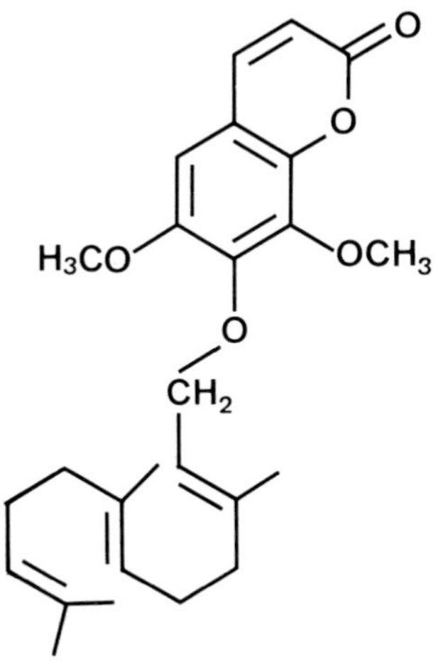

H3CO OCH3 O CH2 O O

XII

H3CO OCH3 O CH2 HO H O O

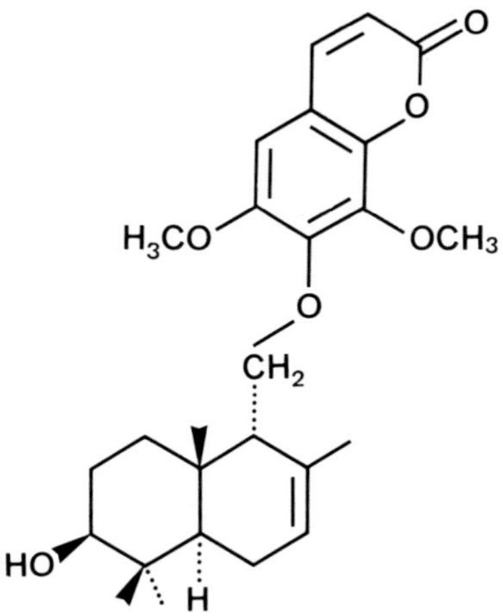

XIII

as a main component from *A. persica, A. gmelinii (=A. iwayomogi* Kitam.), and *A. molinieri* Quezel. In smaller quantities it has also been found in *A. vestita*. The open-chain sesquiterpene ether farnochrol XII is predominantly accumulated in *A. vestita,* but has been detected to some extent in all species mentioned above. It is probably the biogenetic precursor of all derivatives belonging to that class of compounds.

This common chemical trend again supports the close relationship between *A. persica* and the section Abrotanum. Furthermore, from *A. alba* Turra another new coumarin-sesquiterpene ether has been isolated in large quantities which has proven to be an exomethylene isomer of drimartol A XIII (H.Greger & O.Hofer, unpublished data). Taking into account the uncertain taxonomic position of the *A. alba* complex (with "absinthoid" and "abrotanoid" representatives) this biogenetic capacity suggests close affinities with the section Abrotanum.

2.3. Tetrahydrofurofuran lignans

Sesamin-type lignans represent a group of phenylpropane dimers containing a diagonally substituted tetrahydrofurofuran nucleus with four asymmetric centers which may lead to three possible (+/-)-pairs of stereoisomers. The formation of these lignans separates *A. absinthium* L. and the seven closely related species *A. arborescens* L., *A. argentea* L'Herit, *A. canariensis* Less., *A. gorgonum* Webb, *A. siversiana* Willd., *A. macrocephala* Jacquem. ex Bess., and *A. jacutica* Drob. sharply from all the other *Artemisia* species investigated so far (10, 12, 13). All derivatives proved to be components of the resins which accumulate in schizogenous secretory canals.

From this *A. absinthium* group 14 stereoisomers which derived from seven different basic structures could be isolated, all of them with a positive optical rotation. Four of the main components have been proven to be new and were designated as (+)-sesartemin XIV, (+)-episesartemin A XV, (+)-episesartemin B XVI, and (+)-diasesartemin XVII. Within the *A. absinthium* group the different basic structures as well as the stereo-specific chemical differences contribute to a taxonomic grouping which corresponds well with morphological characters and geographical distribution (12).

The common trend towards equatorially/axially (i.e. XV, XVI) and diaxially configurated isomers (i.e. XVII) points to a close affinity between the perennial Eurasiatic *A. absinthium* and the (annual to) biennial *A. siversiana* from Asia. The remaining annual to biennial Asiatic species *A.*

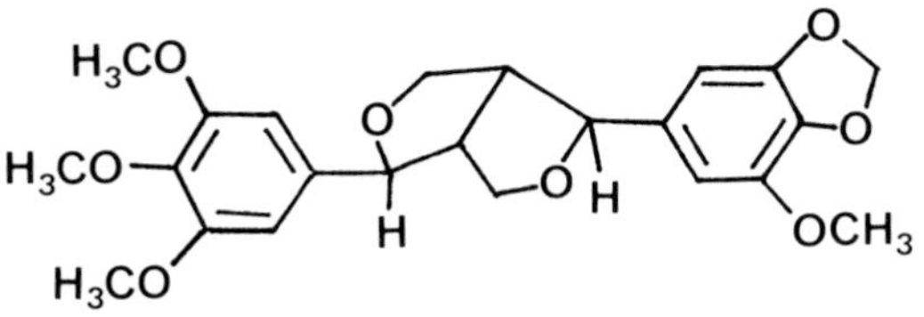

XIV

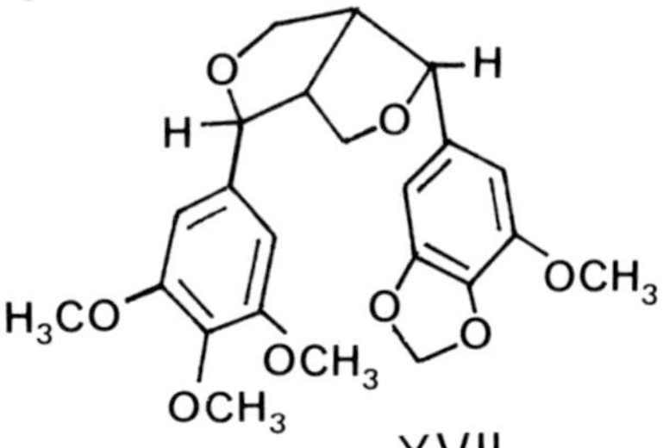

XVII

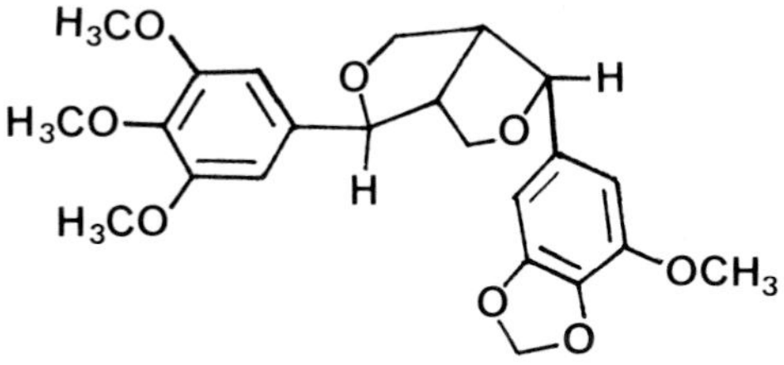

XV

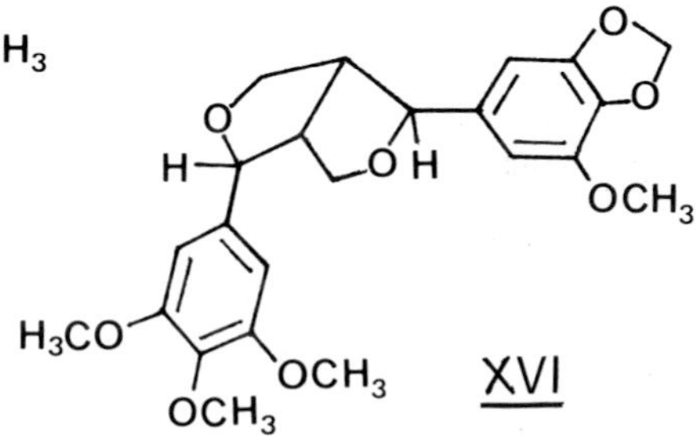

XVI

jacutica and *A. macrocephala* deviate by having a preponderance of diequatorially configurated derivatives (i.e. XIV) and a reduced set of basic structures. Finally, the shrubby members of the Mediterranean (*A. arborescens*) and Macaronesian region (*A. canariensis, A. gorgonum*) are characterized by a more complex lignan pattern with predominately diequatorial isomers (12).

3. CONCLUSION

Summarizing, it is evident from the data presented that chemical characters can play a prominent role in the infrageneric classification of *Artemisia*. Based on a thorough phytochemical screening it became apparent that especially distinct trends in the acetylene biosynthesis serve as valuable systematic criteria. Additionally, the accumulation of sesquiterpene-coumarin ethers as well as tetrahydrofurofuran lignans contribute to a taxonomic grouping within the genus.

Within the presumed "basic stock" of the genus the different trends either towards spiroketalenol ether or pontica epoxide accumulation separate members of the sections Artemisia and Absinthium from those of the section Abrotanum. The subgenera *Dracunculus* and *Seriphidium* are clearly distinguished from this group either by the consistent occurrence of dehydrofalcarinone derivatives together with characteristic aromatic acetylenes in the former or by a general decrease of acetylene production in the latter. Both are recognized as more advanced groups on the basis of floral characters.

The "Heterophyllae" group also deviates from the section Abrotanum by the accumulation of dehydrofalcarinone derivatives as well as by some morphological features, indicating a close alliance with *A. norvegica*. The E-Asiatic *A. keiskeana* can be distinguished from the section Abrotanum by the same chemical trend but, on the basis of inflorescence and involucre morphology, it is more closely allied with the subgenus *Dracunculus*.

The Eurasiatic species of the subgenus *Seriphidium* are characterized by very small amounts of acetylenes lacking spiroketalenol ethers and thus can be separated from the N-American "Tridentatae" with trends towards spiroketalenol ether accumulation.

Ethers with open-chain or bicyclic sesquiterpenes linked to isofraxidin represent a typical chemical trend within the section Abrotanum and point to a close relationship with the taxonomically uncertainly placed *A. alba*. In view of this chemical capacity as well as the accumulation of pontica epoxide, *A. persica* clearly fits in with this group and hence should be excluded from the section Absinthium.

Finally, the formation of sesamin-type lignans sharply separates *A. absinthium* and seven closely related species from the other members of the section Absinthium. In this case, various basic structures as well as stereospecific chemical differences lead to a species grouping which corresponds well with morphological characters and geographical distribution.

REFERENCES

1. Banthorpe DV, Baxendale D, Gatford C, Williams SR. 1971. *Planta Med.* 20: 147.
2. Bohlmann F, Burkhardt T, Zdero C. 1973. Naturally occurring acetylenes. London, Academic Press.
3. Bohlmann F, Ehlers D. 1977. *Phytochemistry* 16:1450.
4. Bohlmann F, Schumann D, Zdero C. 1974. *Chem. Ber.* 107:644.
5. Bohlmann F, Zdero C. 1972. *Phytochemistry* 11:2329.
6. Candolle AP de. 1838. Prodromus systematis naturalis vegetabilis. vol. 6. Paris.
7. Ehrendorfer F. 1964. *Österr. Bot. Z.* 111:84.
8. Epstein WW, Poulter CD. 1973. *Phytochemistry* 12:737.

9. Greger H. 1977. In: The Biology and Chemistry of the Compositae, Heywood VH, Harborne JB, Turner BL (eds.) vol. 2, p. 899. London, Academic Press
10. Greger H. 1979. *Planta Med.* 35:84.
11. Greger H. 1979. *Phytochemistry* 18:1319.
12. Greger H. 1981. *Biochem. Syst. Ecol.* 9:165.
13. Greger H, Hofer O. 1980. *Tetrahedron* 36:3551.
14. Greger H, Hofer O, Nikiforov A. 1982, in press. *J. Nat. Prod.*
15. Hall HM, Clements FE. 1923. The phylogenetic method in taxonomy. The North American species of *Artemisia, Chrysothamnus,* and *Atriplex*. Publ. Carnegie Inst. 326.
16. Harborne JB, Mabry TJ, Mabry H. (eds.) 1975. The flavonoids. London, Chapman and Hall.
17. Hegnauer R. 1973. Chemotaxonomie der Pflanzen. vol. 6. Basel, Birkhäuser Verlag.
18. Heywood VH, Humphries CJ. 1977. In: The Biology and Chemistry of the Compositae, Heywood VH, Harborne JB, Turner BL (eds.) vol. 2, p. 851 London, Academic Press.
19. Hultén E. 1954. *Nytt Mag. Bot.* 3:63.
20. Kelsey RG, Shafizadeh F. 1979. *Phytochemistry* 18:1591.
21. Kraševinnikov IM. 1946. *Mat. ist. fl. rast. SSSR* 2:87.
22. McArthur ED, Pope CL, Freeman DC. 1981. *Amer. J. Bot.* 68:589.
23. Poljakov PP. 1961. In: Flora SSSR, Komarov VL (ed.) vol.26. Moskva, Leningrad.
24. Rydberg A. 1916. In: North American Flora, vol.34. New York.
25. Saidhodžaev AI. 1979. *Khim. Prir. Soedin.*:437.
26. Stangl R, Greger H. 1980. *Pl. Syst. Evol.* 136:125.

CHEMICAL INVESTIGATIONS OF ESSENTIAL OILS OF UMBELLIFERS*

K.-H. KUBECZKA

1. INTRODUCTION

The large family Umbelliferae (=Apiaceae) is rich in secondary metabolites and embodies numerous genera of high economic and medicinal value, yielding coumarins, flavonoids, acetylenic compounds, sesquiterpenic lactones and last but not least essential oils. It is well known that occurrence of essential oils and oleoresins is a characteristic feature of this family but only about 10% of the known species have been investigated for these constituents. The fundamental advancement in the field of essential oils, which was hampered for a long time due to the lack of efficient separation methods, has now been overcome by the application of highly effective chromatographic methods and the modern physical methods of structure elucidation. As a result, our knowledge of the chemistry of volatile compounds is increasing at an enormous rate. Scanning of literature has pointed out that about 760 different constituents have been isolated from essential oils of Umbellifers, belonging to different chemical classes (Table 1).

The present contribution is devoted to the analysis of essential oils from Umbellifers belonging to the subfamily Apioideae, which have not been investigated so far. They will be discussed in succession using grouping into tribes and subtribes according to Drude (2).

2. ANALYTICAL METHODS

Gas chromatography (GLC), especially capillary GLC coupled with mass spectrometry, has become the most important analytical technique in the last decade for separating and analysing essential oils. However, some-

* This paper is based to a large extent on work carried out in the author's department; members of the research team were A. Bartsch, Dr. V. Formacek J. Schwanbeck, I. Ullmann and A. Viernickkel whose contributions are gratefully acknowledged.

Margaris N, Koedam A, and Vokou D (eds.): Aromatic Plants: Basic and Applied Aspects
© 1982. Martinus Nijhoff Publishers, The Hague/Boston/London. ISBN 90-247-2720-0.
Printed in the Netherlands.

Table 1. Number of compounds, isolated from essential oils of Umbellifers (till 1980) (excl. sesquiterpene lactones and coumarins).

Compound class	Number
monoterpenes	159
sesquiterpenes	135
terpene aldehyde esters*	19
phenylpropane derivatives	58
non-terpenic, aliphatic compounds	193
acetylenic compounds	80
phthalides and related compounds	35
sulfur containing compounds	11
nitrogen containing compounds	15
other compounds	54
	759

*not present in steam distillates

times problems can arise which are associated with the thermally instable compounds, the detection and isolation of low concentrations and the need of further spectroscopic data for structure elucidation of unknown new oil constituents. We have overcome these difficulties by application of high performance liquid chromatography (HPLC) using a combination of reverse phase and adsorption chromatography.

2.1. High performance liquid chromatography (HPLC)

By means of reverse phase chromatography an octadecylsilica (LiChrosorb RP 18) column and various proportions of methanol and water a semi-preparative pre-separation of an essential oil into 3 fractions, containing oxygenated compounds, monoterpene hydrocarbons and sesquiterpene hydrocarbons has been obtained (7). The separation of the hydrocarbons was completed by a high performance liquid chromatographic separation step with silica gel/n-pentane at $-15^{o}C$, which provided in many cases pure individual components whose structures could be determined by UV-, IR-, and NMR-spectroscopy (8).

2.2. C-13 NMR spectroscopy

In addition to this, C-13 NMR spectroscopy - usually used for identification and structure elucidation of isolated compounds - was applied to the analysis of essential oils. This new technique, which is a useful tool for direct qualitative and quantitative analysis of total essential oils (3, 4, 10) allows to investigate even complex mixtures without preliminary separation of their components. This is enabled by good separation of the individual NMR signals. The unequivalent carbon atoms of a single component tend to give rise to individually resolved signals under conditions of total proton decoupling. This allows qualitative analysis of the more or less complicated mixture by comparing signal position to the spectra of references. The spectra of both, the essential oil and of reference compounds have to be recorded under identical conditions, to assure no, or only little differences in the chemical shift for the individual C-13 NMR lines of the reference compound and the component of the essential oil. Thus, the results obtained from C-13 NMR spectroscopy are in good accordance with the chromatographic-mass spectrometric findings. Main advantages of the C-13 NMR method are that even non-volatile and thermally unstable components can be analysed without difficulty.

With aid of the mentioned methods numerous oil constituents belonging to the mono- and sesquiterpenoids, norisoprenoids, phenyl propane derivatives, aliphatic- and aromatic compounds have been identified, the majority of which were common to Umbellifers.In addition to these, some hitherto unknown compounds have been isolated and their structure elucidated; these findings will be discussed briefly.

3. RESULTS

3.1. Tribe 1. Echinophoreae

3.1.1. *Echinophora spinosa* L. *Echinophora spinosa* is a perennial plant from the Mediterranean region, growing mainly on maritime sands. The analysis of the essential root oil of this species revealed the presence of different monoterpenes, of an aromatic and of an acetylenic compound; fifteen out of these constituents have been identified. The most significant feature in this oil is the occurrence of 77.2% terpinolene, 4.3% limonene, 9.1% myristicine, and 2.9% falcarinol, a widespread acetylenic compound in Umbellifers. The remaining common monoterpene hydrocarbons are present in just negligible amounts.

The essential oil of the above ground parts showed a different chemical composition: the main constituents are the monoterpene hydrocarbons α-phellandrene (36.8%), *p*-cymene (27.3%), α-pinene (15.0%), β-phellandrene (6.9%), and limonene (2.6%). In addition to these, a hydrocarbon $C_{10}H_{14}$, MW 134 in concentration of about 5% has been isolated from this oil. The structure of this constituent was established spectroscopically as 2,6-dimethylocta-1,3,5,7-all-*trans*-tetraene, a component known to occur in

Cosmos bipinnatus Cav. and some other genera of the family of Compositae (11). This is the first report of the occurrence of cosmene as a constituent of an essential oil of an Umbellifer.

3.2. Tribe 2. Scandiceae

3.2.1. *Chaerophyllum bulbosum* L. From *Chaerophyllum bulbosum,* a biennial or perennial plant, native to eastern and central Europe, with a short tuberous root, we have investigated the essential root- and fruit-oils. Both contain more than 90% monoterpene hydrocarbons, which are common constituents of essential oils, but differ considerably in the quantitative proportions of the single constituents. The main component of the fruit-oil is *trans*-ocimene (85.2%). Minor components are *cis*-ocimene (5.4%), γ-terpinene (2.4%) and β-pinene (1.8%), terpinolene (0.7%), α-pinene (0.6%), myrcene (0.5%), limonene (0.2%), sabinene (0.1%). Besides these constituents, camphene, 3-carene, terpinolene, α- and β- phellandrene are present in traces.

In contrast to the fruit oil the essential root-oil of *Ch. bulbosum* contains no predominating component. Characteristic constituents are *cis*-ocimene (27.3%), *trans*-ocimene (17.6%), limonene (13.5%), α-pinene (12.4%), γ-terpinene (10.0%), and *p*-cymene (8.2%). The remaining components, present in the fruit oil are found in concentrations below 1%.

3.2.2. *Scandix pecten-veneris* L. The genus *Scandix* contains about 15 species mostly confined to the Mediterranean region. All species occur in open habitats, often as weeds but have become now rather rare. The species *S. pecten-veneris* shows a complex range of variations and has been divided into different subspecies from which we have investigated the ssp. *brachycarpa* (Guss.) Thell., and ssp. *pecten-veneris*.

Scandix pecten-veneris ssp. *brachycarpa*, native to Greece and Italy, yielded after steam distillation of the herb a mixture of volatiles consisting mainly of non-terpenic, aliphatic compounds such as the n-alkanes C-13 (19.4%), C-15 (18.5%) and C-17 (23.0%) besides myristic aldehyde (8.0%). In addition to these, we have isolated a component $C_{13}H_{16}O_2$, MW 204, present in a concentration of 9.9%. On the basis of the IR-, MS-, Proton- and C-13 NMR-spectra, this substance has been identified as butyric acid ester of anol, the 4-propenyl phenol. The structure of this component, which has been found for the first time as a constituent of an Umbellifer, was confirmed by synthesis.

Furthermore, we identified minor amounts of corresponding isobutyrate (0.64%), 2-methylbutyrate (less than 0.1%) and 3-methylbutyrate (0.6%).

All these components are also present in the oil of the roots, but the concentrations differ considerably. The root oil contains mainly (53.1%) anylbutyrate.

Scandix pecten-veneris ssp. *pecten-veneris*. The root oil of this subspecies has a similar composition to the above mentioned ssp. *brachycarpa*, but the percentages of the single constituents differ significantly. Main component of this oil is anylisobutyrate (42.8%).

3.2.3. *Scandix australis* L. This species yields essential oils, containing the n-alkanes C-13, C-15, C-17, amyl-butyrate (root 50.1%; herb 12.1%), -isobutyrate, -2-methylbutyrate, -3-methylbutyrate, and -capronate. In addition to these constituents 20.2% dillapiol is present in the herb oil, whereas the roots contain only 0.2% of this phenolether.

3.2.4. *Molopospermum peloponnesiacum* (L.) Koch. *Molopospermum* is a perennial, big plant from the mountains and subalpine zones of the southern Alps and the Pyrenees. The plant has a distinct odour and contains in contrast to the *Scandix* species a large amount of volatiles. The analysis of the root oils from the alpine plants (9) revealed 13 well-known terpenoids (Table 2, TAE*-type). The main component was the monoterpene hydrocarbon 3-carene (35-45% of the total oil). In addition to this compounds we detected two very similar substances with the same formula $C_{10}H_{12}O$, MW 148, which could not be completely identified by mass spectrometry. After isolation their structures were elucidated by UV-, IR- and NMR-spectroscopy as 2,3,4-trimethylbenzaldehyde and 2,3,6-trimethylbenzaldehyde, respectively.

A comparison of the composition of the essential root oil from *Molopospermum* with oils of other Umbellifers reveals significant differences. Only few of them contain 3-carene in significant amounts. The aromatic aldehydes have not yet been detected in essential oils. However, Bohlmann and Zdero (1) obtained 2,3,4-trimethylbenzaldehyde from a ferulol ester upon acid treatment. Therefore it had to be considered that this compound and the isomeric 2,3,6-trimethylbenzaldehyde are produced during steam distillation from corresponding ferulol- and isoferulol derivatives. A comparison of a freshly prepared extract and the distillate from *Molopospermum* roots has finally confirmed this assumption. The aldehydes were not detected in fresh extracts, whereas the steam distillate of this extract contained a significant amount of both components. These aldehydes have to be considered therefore as artefacts, produced during steam distillation from ferulol-

and isoferulol esters by saponification and proton catalyzed rearrangements, as illustrated below.

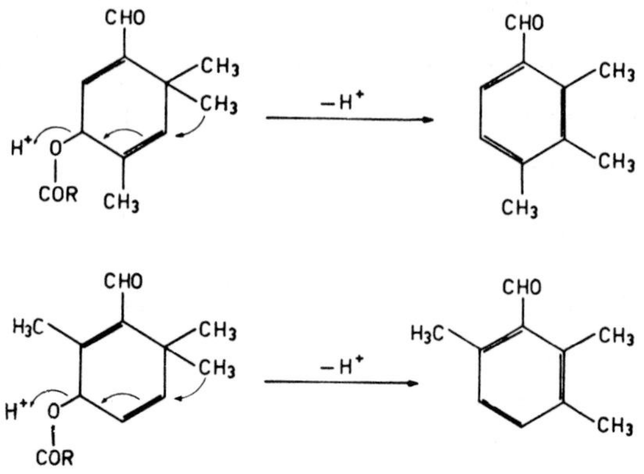

Since ferulol and isoferulol esters have been found in several genera of Umbellifers, it has to be expected that the aromatic aldehydes occur in steam distillates of numerous species. We have detected them in the essential oils from *Anthriscus cerefolium*, *Ligusticum lucidum*, *Selinum carvifolia* and *Silaum silaus* in considerably high amounts. Since these species belong to different tribes of the Umbellifers, the occurrence of these constituents seems to possess no chemosystematic significance.

The investigation of *Molopospermum* roots from the botanical garden Wuppertal (Germany) afforded an essential oil with a significant different composition (Table 2, Apiol-type).

The main constituent of this type of oil is dillapiol (66.7%) which is present in the former type only in traces. In addition to this phenolether small amounts of myristicine and a third phenylpropane derivative have been detected. The structure of the latter constituent was established by spectroscopic methods as 2,5,6-trimethoxy-3,4-methylenedioxy allylbenzene. This component, known under the name nothapiol is a main constituent of the fruit oil of this type. Comparing the compositions of the two types - the second one has also been found in a wild growing plant from the Pyrenees - we can distinguish between an apiol-type, which seems to occur only in the western part of Europe and a terpene aldehyde ester-type from the central Alps. The pattern of monoterpene hydrocarbons is, within certain limitations, in both types relatively similar and is characterized by a high percentage of 3-carene.

Table 2. Components of the essential oil types from *Molopospermum peloponnesiacum* roots.

No.	Compounds	percentages Apiol-type	TAE*-type
1	α-Pinene	3.7	2.5
2	Camphene	< 0.1	< 0.1
3	β-Pinene	0.4	< 0.1
4	Sabinene	< 0.1	0.2
5	3-Carene	15.8	36.4
6	Myrcene	0.6	1.8
7	α-Phellandrene	0.4	0.4
8	α-Terpinene	0.1	< 0.1
9	Limonene	1.0	2.0
10	β-Phellandrene	1.5	1.7
11	*cis*-Ocimene	0.2	-
12	γ-Terpinene	0.2	0.2
13	*trans*-Ocimene	< 0.1	< 0.1
14	*p*-Cymene	< 0.1	0.2
16	Terpinolene	1.2	1.9
18	Thymolmethylether	-	0.4
23	β-Selinene	-	3.5
24	β-Bisabolene	-	0.5
29	2,3,4-Trimethylbenzaldehyde	0.9	9.3
31	2,3,6-Trimethylbenzaldehyde	0.5	27.9
34	Myristicine	0.3	-
35	Dillapiol	66.7	0.1
37	Nothapiol	0.5	-
	Total	95.4	86.9

*Terpene-Aldehyde-Ester-type

3.2.5. *Myrrhis odorata* (L.) Scop. *Myrrhis odorata* is a native of the southern European mountains from the Pyrenees to the western part of the Balkan peninsula. Due to its use in folk medicine and as an aromatic spice the plant is widely naturalized elsewhere. We have analyzed all parts of this plant and confirmed the presence of anethole and smaller amounts of estragole in all of them. Besides these components, which are responsible for the anise-like odour of this plant, the herb oil with 76.8% anethole and 1.2% estragole contains remarkable amounts of terpene hydrocarbons such as α-terpinene (2.4%), limonene (3.5%), β-caryophyllene (3.9%), germacrene-D (4.4%), α-farnesene (2.4%), most of which are also present, though in lower concentrations, in the root oil. But the root oil differs only quantitatively from the oils of the other parts of this plant and contains in addition to these higher amounts of tetradecanol-1 (3.2%), hexadecanol-1 (0.7%) and two different phenylpropane derivatives, whose structures were established as 2,4-dimethoxyallylbenzene (1.4%) and the propenylic isomer (12.3%).

The first of these constituents, known under the name osmorhizol has been isolated previously from *Osmorhiza japonica* (5) and the garden chervil, *Anthriscus cerefolium* (12), all belonging to the same tribe. This component seems therefore to possess a taxon specific distribution.

3.3. Tribe 5. Apieae

3.3.1. *Bupleurum falcatum* L.

3.3.2. *Falcaria vulgaris* Bernh.

The essential oils of the herb and fruit of these two species, native to Europe and western Asia, are remarkable, since they contain more than 50% of the labile sesquiterpene hydrocarbon germacrene-D (6), which is a key intermediate in sesquiterpene biosynthesis. Similar high amounts of this component have only been isolated from one species of the Araliaceae, showing the close chemical relationship of these two families. We are sure to find germacrene-D in further essential oils of Umbellifers, if thermal rearrangements during the analytical procedure are avoided and sensitive analytical techniques are applied.

The aim of this communication was to point out the differences and similarities to be found in the different species of Umbelliferae, which are still today a rich source of secondary plant products.

REFERENCES

1. Bohlmann F, Zdero C. 1969. *Chem. Ber.* 102:2211
2. Drude O. 1898. In: Die Natürlichen Pflanzenfamilien, Engler A, Prantl K (eds.). Vol. 3. Leipzig, Engelmann.
3. Formacek V, Kubeczka K-H. 1979. In: Vorkommen und Analytik ätherischer Öle, Kubeczka K-H (ed.) p. 130. Stuttgart, Thieme Verlag.
4. Formacek V, Kubeczka K-H. 1981. Essential Oil Analysis by GC and C-13 NMR Spectroscopy. London, Heyden & Son.

5. Konoshima M, Hata K, Ikeshira Y. 1967. *Yakugaku Zasshi* 87:1138.
6. Kubeczka K-H. 1979. *Phytochemistry* 18:1066.
7. Kubeczka K-H, Formacek V, Schwanbeck J. 1980. New Techniques in Essential oil analysis. 8th Int. Congress of Essential Oils, Cannes (France).
8. Kubeczka K-H, Schwanbeck J. 1977. *Planta Med.* 32A:39.
9. Kubeczka K-H, Ullmann I. 1981. *Phytochemistry* 20:828.
10. Kubeczka K-H, Ullmann I. 1981. *Rivista Italiana EPPOS* 63:265.
11. Sörensen NA, Sörensen JS. 1954. *Acta Chem. Scand.* 8:284.
12. Zwaving JH, Smith D, Bos R. 1971. *Pharm. Weekbl.* 106:182.

CHAPTER 4
Analysis and Composition

^{13}C-NMR ANALYSIS OF ESSENTIAL OILS

V. FORMACEK, K.-H. KUBECZKA

1. INTRODUCTION

There is little doubt that ^{13}C-NMR spectroscopy is today one of the most important methods in instrumental analysis, finding a wide variety of applications in chemistry, physics, and biology. ^{13}C-NMR spectroscopy is generally used for the elucidation of the molecular dynamics and the determination of constitution, configuration and conformation of polymers and biopolymers. Generally such applications focus on the study of isolated chemical species. The application of ^{13}C-NMR spectroscopy to the investigation of complex mixtures is relatively seldom.

For the analysis of the essential oils ^{13}C-NMR spectroscopy offers particular advantages for solving certain interesting problems more easily and successfully than is possible with other physical or physicochemical techniques. Most important is the ability to analyse oil samples without preliminary separation of their components, made feasible by good separation of the individual NMR signals. Furthermore, direct information about molecular structure and functional groups of individual components is obtained from chemical shift values.

For routine analysis of the essential oils the proton broad-band decoupling technique is of most practical use because of its sensitivity and selectivity. Only in special cases other techniques such as off-resonance decoupling, gated decoupling or selective excitation are recommended.

The qualitative analysis of the investigated essential oil is based upon the comparison of the oil spectrum with the spectra of the pure oil components, which should be recorded, if possible, under identical conditions (solvent, temperature, lock substance, reference etc.). This assures that differences in the chemical shift for the individual ^{13}C-NMR lines of the reference substance and of the mixture are negligible.

A simple example of the qualitative interpretation of an essential oil ^{13}C-NMR spectrum is shown in Fig. 1. Star anise oil main component is *trans*-anethole (ca 85%). The second component is limonene, the concentration of

Margaris N, Koedam A, and Vokou D (eds.): Aromatic Plants: Basic and Applied Aspects

ISBN 90-247-2720-0.
Printed in the Netherlands.

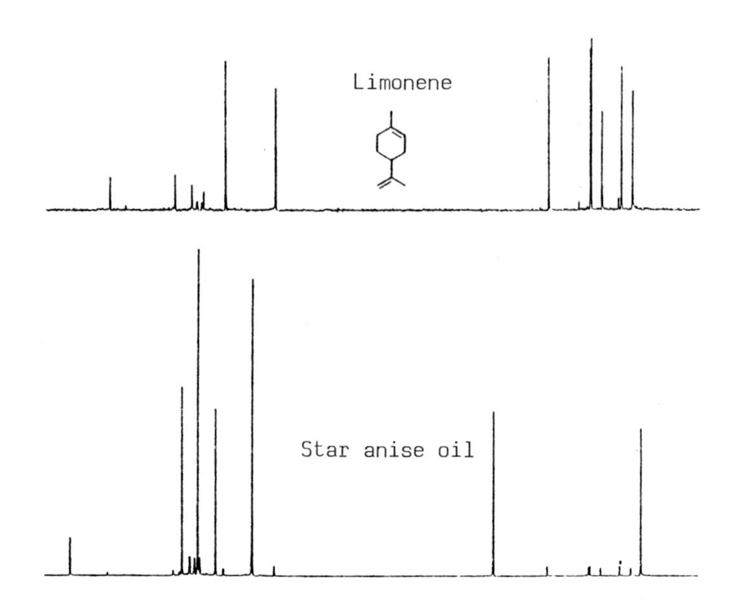

FIGURE 1. ^{13}C-NMR spectra of star anise oil and limonene.

which (ca 8%) gives rise to less intensive resonance signals, which are easy to recognise with help of the reference spectrum.

Figure 2 shows the ^{13}C-NMR spectrum of a more complex mixture, the fennel oil. The main components, *trans*-anethole, fenchone and anise aldehyde can be easily identified with help of the reference spectra as shown above.

In Fig. 2 only the most characteristic signals of the main components are indicated.

Figures 3 and 4 show the expanded parts of the fennel oil ^{13}C-NMR spectrum where the resonance signals of the minor components are indicated.

The sensitivity of the ^{13}C-NMR method is limited by diverse factors (rotational side bands, ^{13}C-^{13}C couplings etc.) and is depending upon the accumulation time. For practical use, the concentration of 0.1% v/v of the terpenic component in the entire mixture is to be seen as an interpretable limit.

For the quantitative analysis of the mixtures using the ^{13}C-NMR proton decoupled spectra, the following error inducing factors, should be considered:

1. Digital resolution

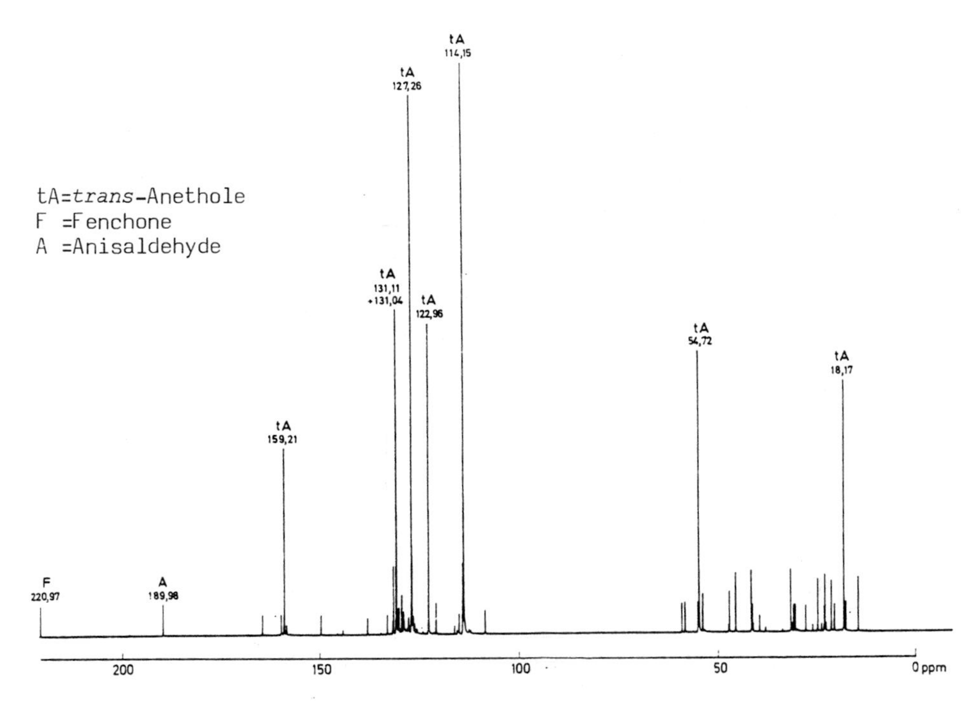

FIGURE 2. ^{13}C-NMR spectrum of fennel oil.

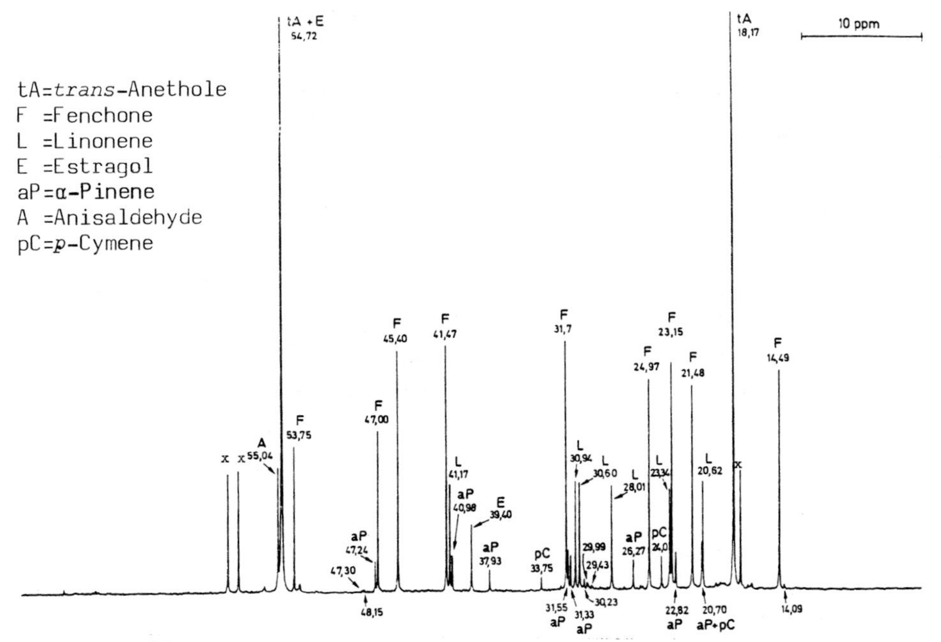

FIGURE 3. ^{13}C-NMR spectrum of fennel oil expanded aromatic region.

2. Relaxation time

3. Nuclear Overhauser Effect

Negative influence of the last two effects can be eliminated by selectioning of the nonprotonated NMR signals. The digital resolution error can be minimized by calculation of the average intensity value.

For the analysis of the essential oils, good quantitative results were achieved by comparison of the average intensity values of the protonated ^{13}C-NMR signals for each of the qualitatively determinated components.

The comparison of the GC and ^{13}C-NMR quantitative analysis of fennel oil is shown in Table 1.

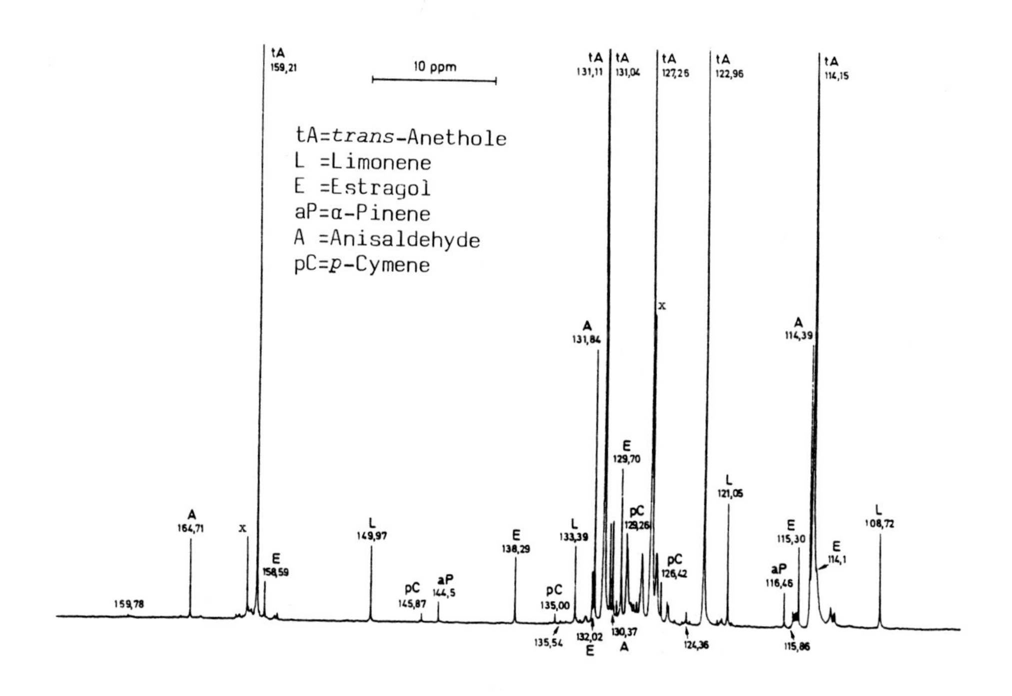

FIGURE 4. ^{13}C-NMR spectrum of fennel oil expanded aliphatic region.

Table 1. Comparison of quantitative analyses of fennel oil by ^{13}C-NMR spectroscopy and gas chromatography.

Compounds	Percentages	
	^{13}C-NMR*	GC†
trans-Anethole	25.6	21.36
Limonene	25.0	26.80
Estragol	2.5	2.07
Fenchone	16.7	14.61
p-Cymene	1.2	1.73
α-Pinene	9.2	10.45
β-Pinene	2.8	3.22
Camphene	0.4	0.49
1.8-Cineole	2.1	2.47
Myrcene	2.5	2.69
α-Phellandrene	9.9	9.72
β-Phellandrene	1.1	1.09
Total	98.6	96.70

* ^{13}C-NMR percentages are based on the average intensity value I of an individual component n (signals of non protonated carbons are eliminated). MW = molecular weight.

$$\%n = \frac{MW \cdot I \cdot 100}{MW_a \cdot I_a + MW_b \cdot I_b \ldots\ldots + MW_n \cdot I_n \ldots + MW_x \cdot I_x}$$

† GC percentages are based on computer calculated area normalization.

SUGGESTED LITERATURE

1. Bellanato J, Hidalgo A. 1971. Infrared analysis of essential oils. London, Heyden &Son.
2. Formacek V, Kubeczka K-H. 1979. Einsatzmöglichkeiten der 13C-NMR Spektro- bei der Analyse ätherischer Öle. In: Vorkommen und Analytik ätherischer Öle, Kubeczka K-H (ed.) p. 130. Stuttgart, Thieme Verlag.
3. Formacek V. 1979. Thesis, Würzburg.
4. Formacek V, Kubeczka K-H. 1981. Essential Oil Analysis by GC and C-13 NMR Spectroscopy. London, Heyden & Son.

QUALITATIVE EVALUATION OF AROMATIC HERBS BY DIRECT HEAD SPACE $(GC)^2$ ANALYSIS. APPLICATIONS OF THE METHOD AND COMPARISON WITH THE TRADITIONAL ANALYSIS OF ESSENTIAL OILS

F. CHIALVA, G. GABRI, P.A.P. LIDDLE, F. ULIAN

1. INTRODUCTION

Head space (HS) analysis is one of the most useful auxiliary techniques available in GC. Its simplicity and rapidity of execution, the possibility of avoiding the laborious treatments required for the preparation of samples, its non-destructive operation which does not modify the sample composition and does not change the structure of the substances under examination, are the most important advantages of this efficient and elegant means of research.

Potential applications of HS method are numerous and easy to imagine. Indeed, all the volatile constituents of certain matrices could, with appropriate preparation, be analyzed by means of the HS technique. In this way it would be possible to examine the substances responsible for the specific olfactory properties of a particular product, by evaluating the constituents of the odour.

In the vast field of aromatisers, and more particularly in the alimentary sector, the use of HS analysis leads to interesting and promising results. This system, while contributing to the solution of particular analytical problems, can also serve as a support for the sensorial analysis which aims for an objective judgment of quality.

The high level of efficiency, the great discriminatory power and the considerable overall sensitivity of the gas chromatograph-detector would therefore provide information which would permit a better definition of the olfactory sensation perceptible during the sensorial examination. Finally, the correct linking of the objective and impartial data of HS analysis with the subjective results of the organoleptic examination, would enable us to obtain indications useful for a more precise and complete judgment of the quality of a raw material.

Margaris N, Koedam A, and Vokou D (eds.): Aromatic Plants: Basic and Applied Aspects

ISBN 90-247-2720-0.
Printed in the Netherlands.

2. EXPERIMENTAL

In a previous study we described a rapid method for the qualitative evaluation of aromatic herbs and spices by means of HS analysis with capillary columns (5). The technique consists of grinding a given quantity of a sample in a blender equipped with a gas-tight valve and submitting to GC analysis the substances present in the gaseous phase, in thermodynamic equilibrium with the herb. The state of equilibrium is reached in time and temperature conditions which are suitably measured. The information obtained is sufficient to establish the fingerprint of the examined herb and is in accordance with the organoleptic judgment expressed by a panel of tasters.

In this study the recognition of the most important volatile substances present in the odour of several aromatic herbs has been achieved; in addition the data obtained from HS analysis have been compared with those of the analysis carried out on the essential oil of the same herb. In this way the process of correlation between the sensorial characteristics, composition and quality is more complete.

In general, the GC analysis of essential oils is commonly carried out to evaluate a particular herb on the basis of its qualitative and quantitative constitution. The analytical values obtained in this way do not, however, refer directly to the volatile substances really present, and therefore to the effective quality of the herb. Indeed, the steam distillation from which the essential oil is obtained, occurs in conditions which can make the composition of the herb variable: the most volatile constituents are usually not recovered, while other components can undergo modifications altering the real picture of the aromatic profile.

We wish to point out again that the essential oils of the herbs examined have been the subject of numerous thorough studies: *Anthemis nobilis* L., roman camomile (6, 7); *Artemisia absinthium* L., common wormwood (1); *Artemisia pontica* L., roman wormwood (3); *Coriandrum sativum* L., coriander (8); *Mentha piperita* Huds., peppermint; *Satureja hortensis* L., savory (4); *Hypericum perforatum* L., St. John's wort and *Teucrium chamaedrys* L., common germander (2). The information concerning the composition and the characterization by HS is, on the other hand very limited.

The herb under examination is finely ground in a glass blender (such as Braun Multimix MX32), and a quantity of 0.5 to 0.1 g, depending on the type of herb, is put in a 10 ml glass vial equipped with a hermetically closing teflon cap. After conditioning at 60°C for one hour in the thermostatic bath of the automatic sampler, the GC head space analysis is carried out.

The HS analyses were performed by means of a gas chromatograph Carlo Erba Model Fractovap 2920 coupled with an automatic sampler Model HS 250. The use of this apparatus, which operates in a completely automatic cycle, enables us to avoid possible errors in sampling, and allows us to obtain results which can be reproduced perfectly.

The conditions adopted are as follows:

column: 25 m, OV 1, 0.15 μm stationary phase thickness
carrier gas: H_2, flow rate 2.1 ml min^{-1}
split ratio: 1:4
injector and detector temperatures: 150°C
oven: 3 min hold at 25°C, 3°C min^{-1} to 180°C
quantity injected: 1 ml, attenuation x 4
sampler syringe temperature: 80°C
sampler bath temperature: 60°C

For the analyses of the essential oils we worked with the same column, flow and programming, but the temperature of the injector and the FID was raised to 200°C and the split ratio was changed to 1:40; quantity injected 0.01 μl, attenuation x 4.

The data were processed by means of a P.E. integrator, model Sigma 10. The analyses in GC-MS were also performed on all the samples by means of HP apparatus model 5992A, operating in EI. The gas chromatograms relating to the head spaces and to the corresponding essential oils of the herbs examined are reported in Figs. 1-8. The substances identified in the essential oil and in the odour of the herbs analysed appear in the attached tables with the relative percentage concentration. These figures do not have an absolute quantitative value but allow to point out the different quantities of some typical compounds present in the essence and in the odour.

3. RESULTS AND DISCUSSION

The gas chromatograms relative to the head space and the essential oil of the various herbs considered, appear clearly differentiated, easily distinguishable and characteristic.

On examining the results obtained as a whole, it can be seen that in the GC plot of the essential oils in general a larger number of sub-

FIGURE 1 *ANTHEMIS NOBILIS L.*

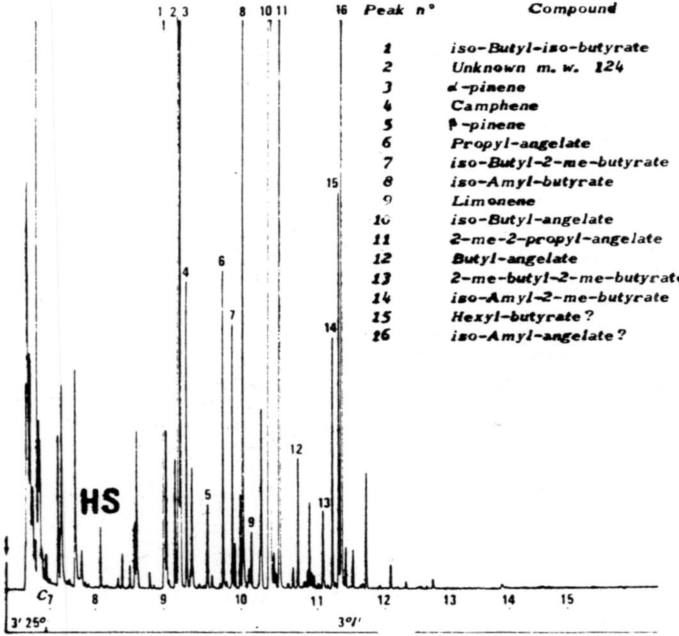

Peak n°	Compound	% OIL	% HS
1	iso-Butyl-iso-butyrate	3,7	12,7
2	Unknown m. w. 124	2,4	8,3
3	α-pinene	1,6	3,7
4	Camphene	0,4	1,1
5	β-pinene	0,2	>0,1
6	Propyl-angelate	1,1	1,1
7	iso-Butyl-2-me-butyrate	0,7	0,8
8	iso-Amyl-butyrate	2,6	3,4
9	Limonene	0,2	>0,1
10	iso-Butyl-angelate	36,0	36,5
11	2-me-2-propyl-angelate	7,4	5,8
12	Butyl-angelate	0,9	0,2
13	2-me-butyl-2-me-butyrate	2,7	>0,1
14	iso-Amyl-2-me-butyrate	2,8	0,8
15	Hexyl-butyrate?	3,9	1,4
16	iso-Amyl-angelate?	17,9	8,4

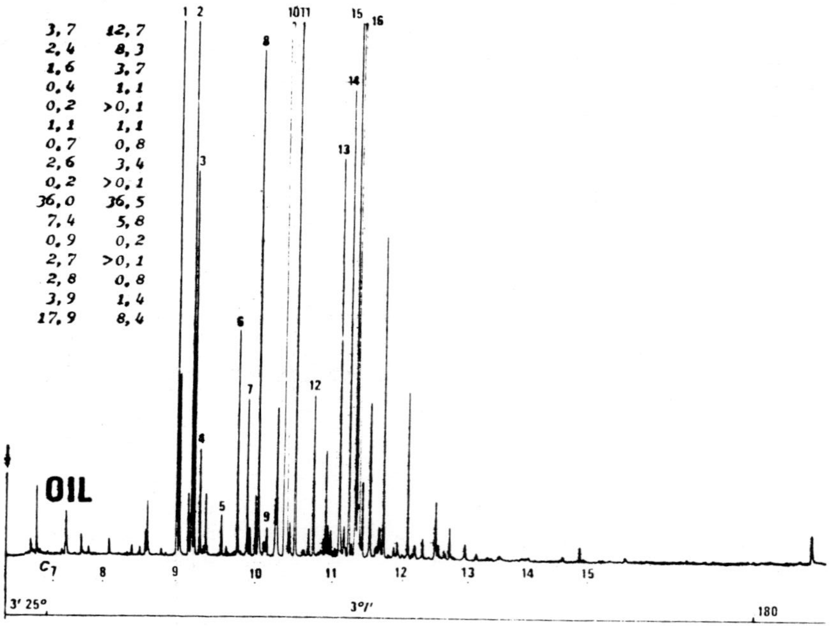

FIGURE 2 ARTEMISIA ABSINTHIUM L.

Peak n°	Compound	% OIL	% HS
1	α-pinene	0,1	2,6
2	Camphene	0,1	0,8
3	β-pinene	0,2	2,6
4	Myrcene	0,2	2,3
5	p-cymene	>0,1	>0,1
6	1,8-cineole	>0,1	>0,1
7	Limonene	>0,1	>0,1
8	γ-terpinene	>0,1	>0,1
9	Linalool	6,0	10,6
10	β-thujone	2,7	2,0
11	cis-Epoxy-ocimene	16,7	10,3
12	trans-Epoxy-ocimene	2,2	0,7
13	Camphor	2,5	1,5
14	Terpinen-4-ol	2,0	6,0
15	Nerol	1,2	
16	Geraniol	3,1	
17	Chrysanthenyl-acetate	4,4	0,6
18	Eugenol	0,4	
19	α-cubebene	0,4	
20	β-bourbonene	0,7	
21	Caryophyllene	1,6	
22	Hydroc. sesquit. m. w. 204	0,8	
23	Unknown m. w. 186	1,1	
24	(Neryl/geranyl)-2-me-butyrate	3,2	
25	(Neryl/geranyl)-2-me-butyrate	2,1	
26	α-bisabolol	1,0	
27	Chamazulene	1,2	
28	Unknown m. w. 248	0,3	

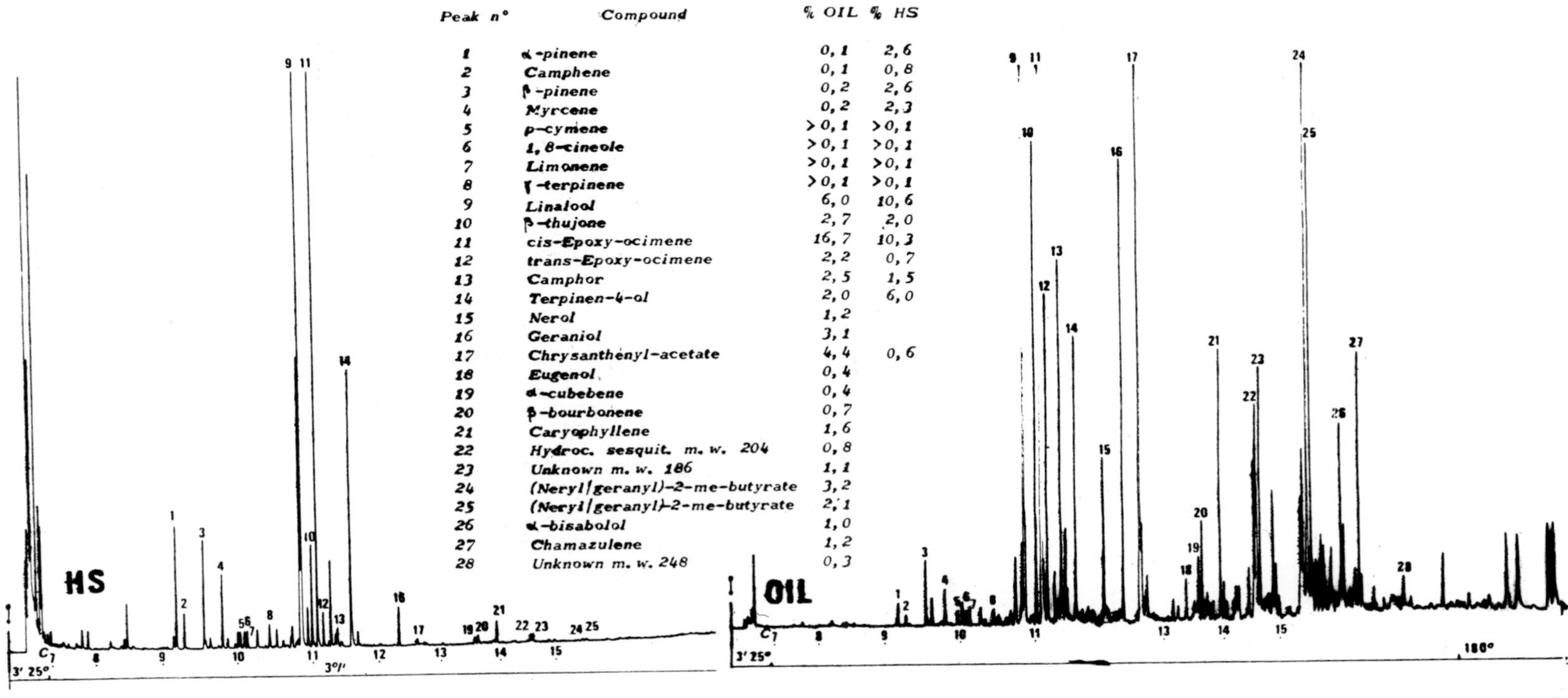

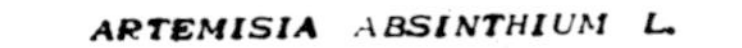

FIGURE 3 ARTEMISIA PONTICA L.

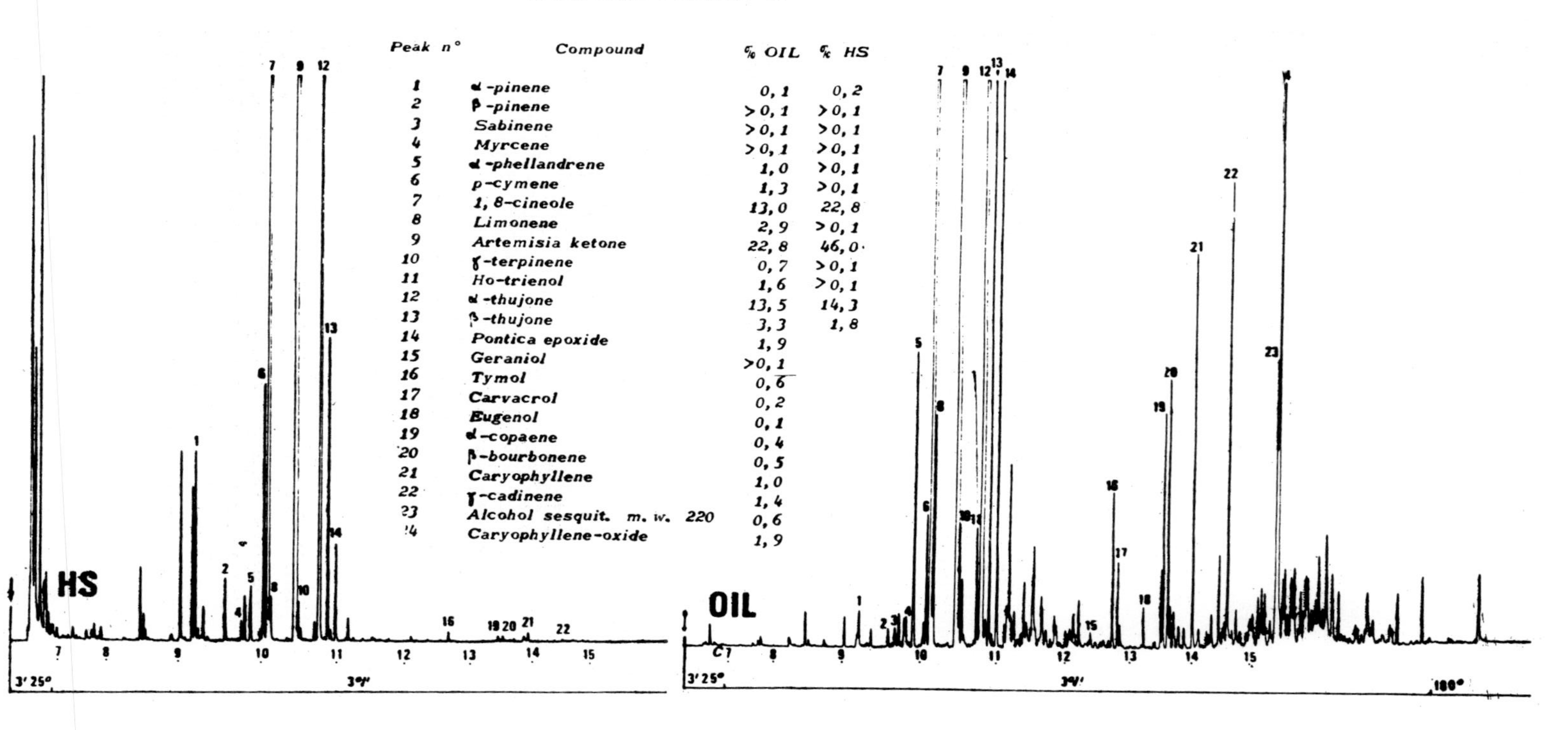

Peak n°	Compound	% OIL	% HS
1	α-pinene	0,1	0,2
2	β-pinene	>0,1	>0,1
3	Sabinene	>0,1	>0,1
4	Myrcene	>0,1	>0,1
5	α-phellandrene	1,0	>0,1
6	p-cymene	1,3	>0,1
7	1,8-cineole	13,0	22,8
8	Limonene	2,9	>0,1
9	Artemisia ketone	22,8	46,0
10	γ-terpinene	0,7	>0,1
11	Ho-trienol	1,6	>0,1
12	α-thujone	13,5	14,3
13	β-thujone	3,3	1,8
14	Pontica epoxide	1,9	
15	Geraniol	>0,1	
16	Tymol	0,6	
17	Carvacrol	0,2	
18	Eugenol	0,1	
19	α-copaene	0,4	
20	β-bourbonene	0,5	
21	Caryophyllene	1,0	
22	γ-cadinene	1,4	
23	Alcohol sesquit. m. w. 220	0,6	
24	Caryophyllene-oxide	1,9	

FIGURE 4 CORIANDRUM SATIVUM L.

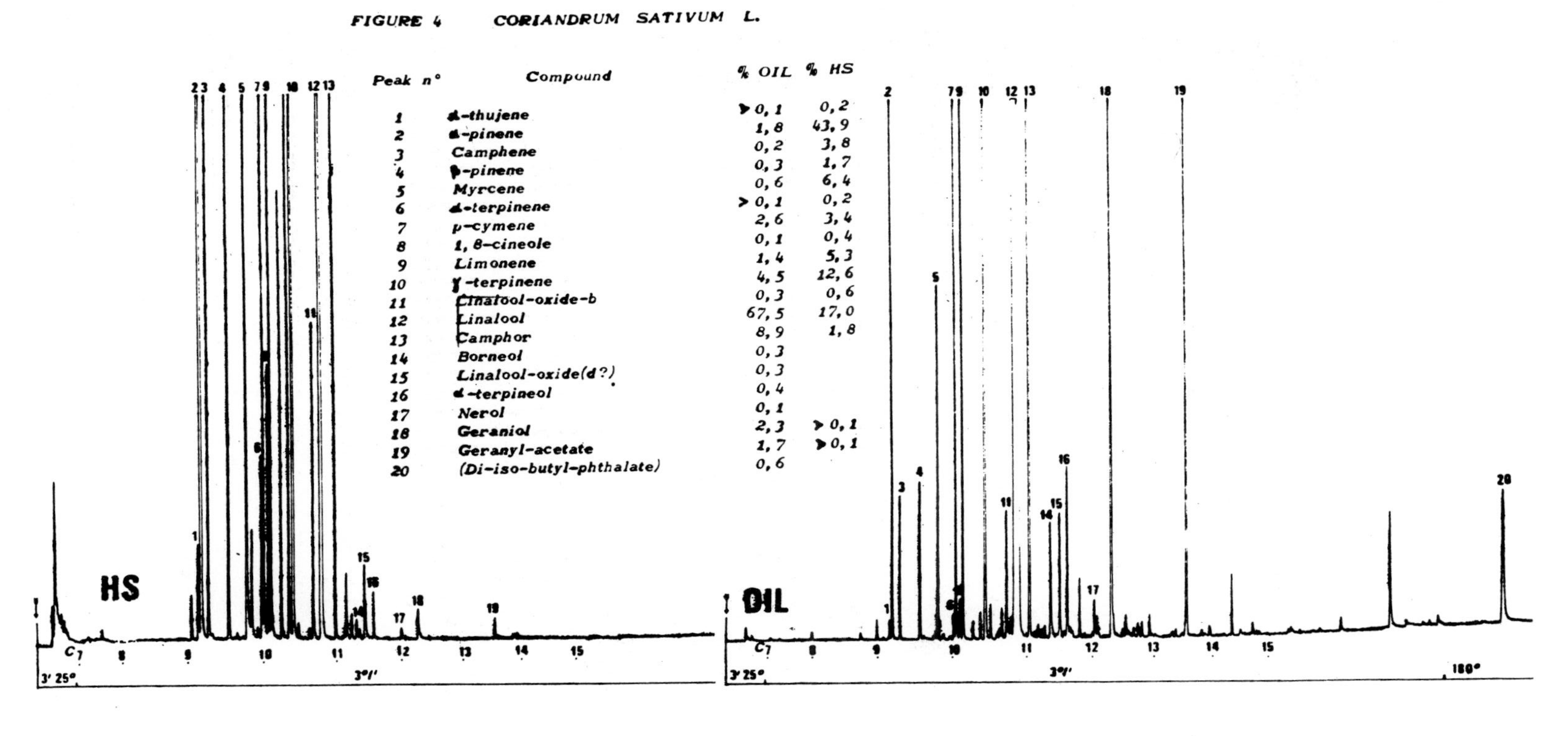

Peak n°	Compound	% OIL	% HS
1	α-thujene	> 0,1	0,2
2	α-pinene	1,8	43,9
3	Camphene	0,2	3,8
4	β-pinene	0,3	1,7
5	Myrcene	0,6	6,4
6	α-terpinene	> 0,1	0,2
7	p-cymene	2,6	3,4
8	1,8-cineole	0,1	0,4
9	Limonene	1,4	5,3
10	γ-terpinene	4,5	12,6
11	Linalool-oxide-b	0,3	0,6
12	Linalool	67,5	17,0
13	Camphor	8,9	1,8
14	Borneol	0,3	
15	Linalool-oxide(d?)	0,3	
16	α-terpineol	0,4	
17	Nerol	0,1	
18	Geraniol	2,3	> 0,1
19	Geranyl-acetate	1,7	> 0,1
20	(Di-iso-butyl-phthalate)	0,6	

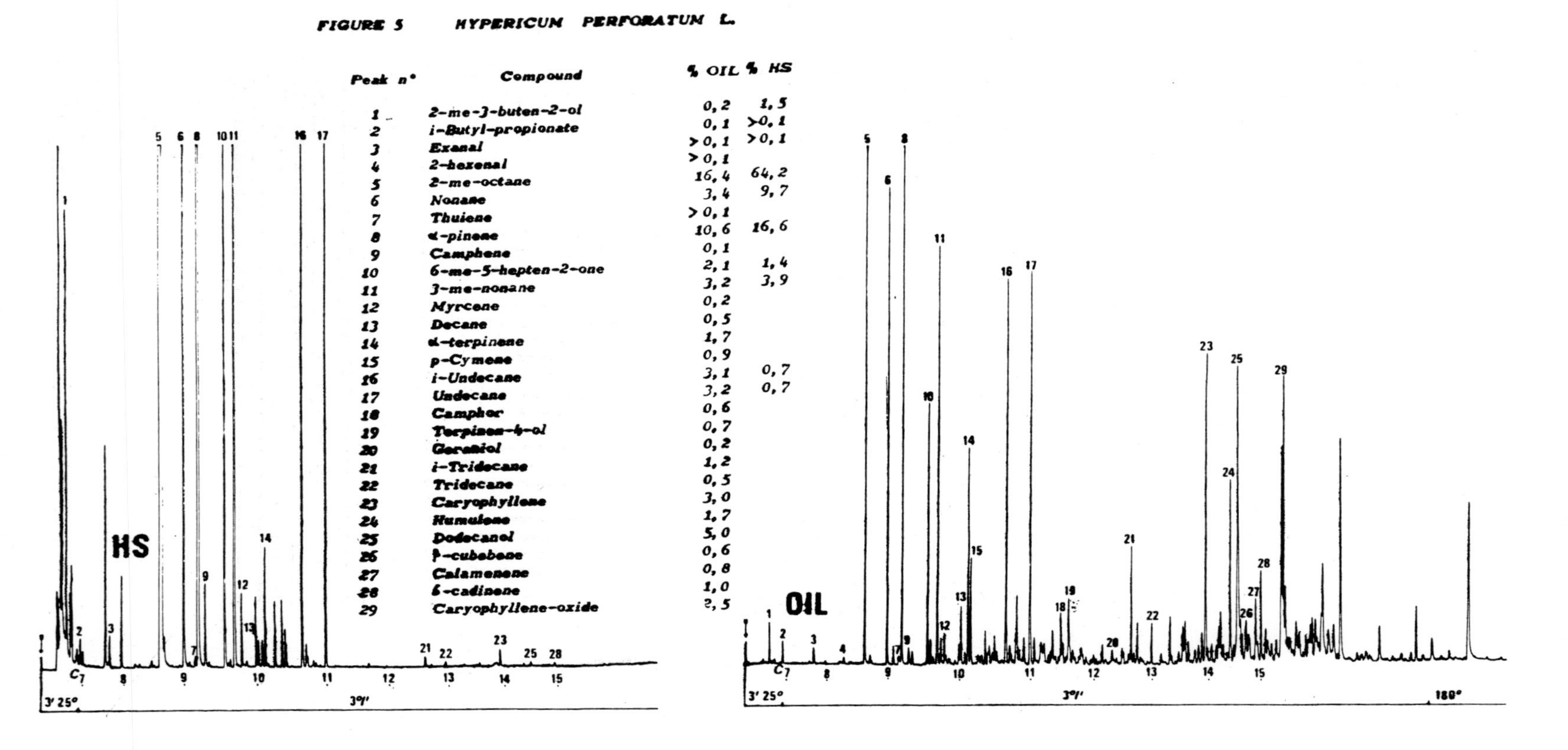

FIGURE 5 HYPERICUM PERFORATUM L.

Peak n°	Compound	% OIL	% HS
1	2-me-3-buten-2-ol	0,2	1,5
2	i-Butyl-propionate	0,1	>0,1
3	Exanal	>0,1	>0,1
4	2-hexenal	>0,1	
5	2-me-octane	16,4	64,2
6	Nonane	3,4	9,7
7	Thuiene	>0,1	
8	α-pinene	10,6	16,6
9	Camphene	0,1	
10	6-me-5-hepten-2-one	2,1	1,4
11	3-me-nonane	3,2	3,9
12	Myrcene	0,2	
13	Decane	0,5	
14	α-terpinene	1,7	
15	p-Cymene	0,9	
16	i-Undecane	3,1	0,7
17	Undecane	3,2	0,7
18	Camphor	0,6	
19	Terpinen-4-ol	0,7	
20	Geraniol	0,2	
21	i-Tridecane	1,2	
22	Tridecane	0,5	
23	Caryophyllene	3,0	
24	Humulene	1,7	
25	Dodecanol	5,0	
26	β-cubebene	0,6	
27	Calamenene	0,8	
28	δ-cadinene	1,0	
29	Caryophyllene-oxide	2,5	

FIGURE 6 MENTHA PIPERITA Huds.

Peak n°	Compound	% OIL	% HS
1	α-thujene	0,1	0,4
2	α-pinene	0,6	5,5
3	Camphene	>0,1	
4	β-pinene	1,3	7,4
5	Sabinene	0,2	
6	Myrcene	0,6	1,2
7	α-terpinene	0,5	1,0
8	p-cymene	0,2	
9	1,8-cineole	5,4	5,3
10	Limonene	1,9	7,2
11	cis-β-ocimene	0,9	1,7
12	γ-terpinene	3,6	2,7
13	Linalool	1,0	
14	Menthone	19,7	23,7
15	iso-Menthone	2,7	3,4
16	Menthofuran	2,9	1,4
17	Menthol	31,1	15,5
18	α-terpineol	1,3	
19	Piperitone	1,1	
20	Menthyl-acetate	1,2	0,2
21	β-bourbonene	0,5	
22	Caryophyllene	1,0	0,2
23	Humulene	1,6	
24	δ-cadinene	0,3	
25	Viridiflorol	0,1	

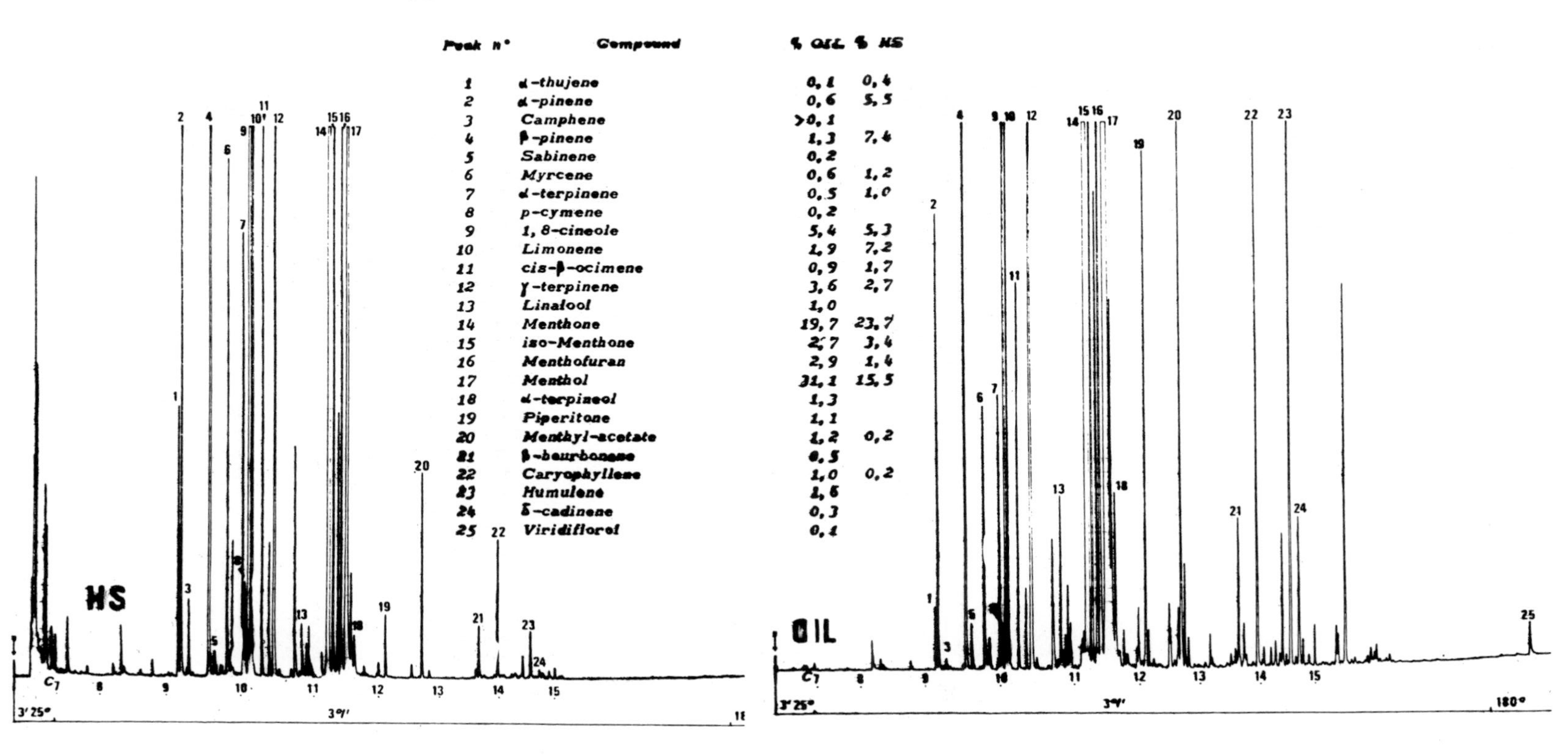

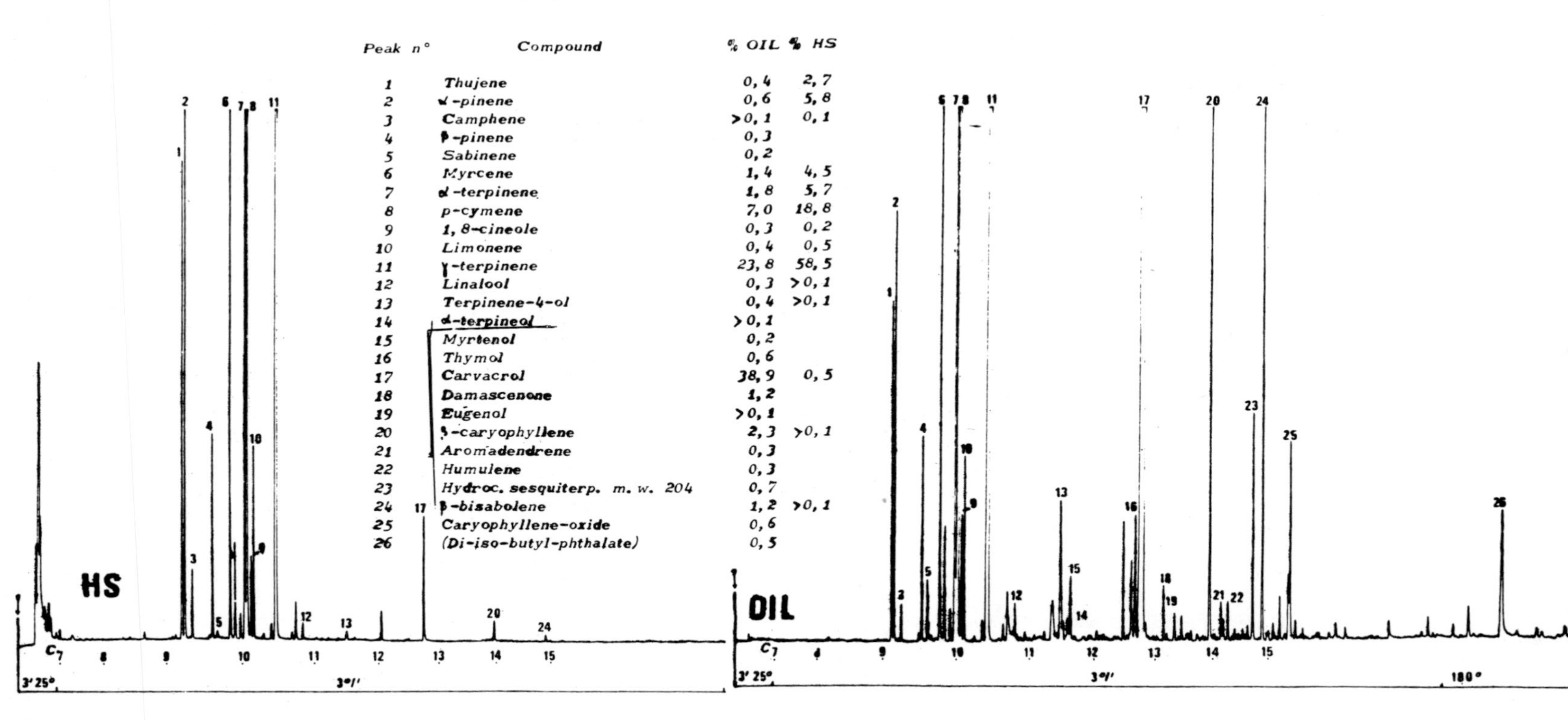

FIGURE 7 SATUREJA HORTENSIS L.

Peak n°	Compound	% OIL	% HS
1	Thujene	0,4	2,7
2	α-pinene	0,6	5,8
3	Camphene	>0,1	0,1
4	β-pinene	0,3	
5	Sabinene	0,2	
6	Myrcene	1,4	4,5
7	α-terpinene	1,8	5,7
8	p-cymene	7,0	18,8
9	1,8-cineole	0,3	0,2
10	Limonene	0,4	0,5
11	γ-terpinene	23,8	58,5
12	Linalool	0,3	>0,1
13	Terpinene-4-ol	0,4	>0,1
14	α-terpineol	>0,1	
15	Myrtenol	0,2	
16	Thymol	0,6	
17	Carvacrol	38,9	0,5
18	Damascenone	1,2	
19	Eugenol	>0,1	
20	β-caryophyllene	2,3	>0,1
21	Aromadendrene	0,3	
22	Humulene	0,3	
23	Hydroc. sesquiterp. m. w. 204	0,7	
24	β-bisabolene	1,2	>0,1
25	Caryophyllene-oxide	0,6	
26	(Di-iso-butyl-phthalate)	0,5	

FIGURE 8 TEUCRIUM CHAMAEDRYS L.

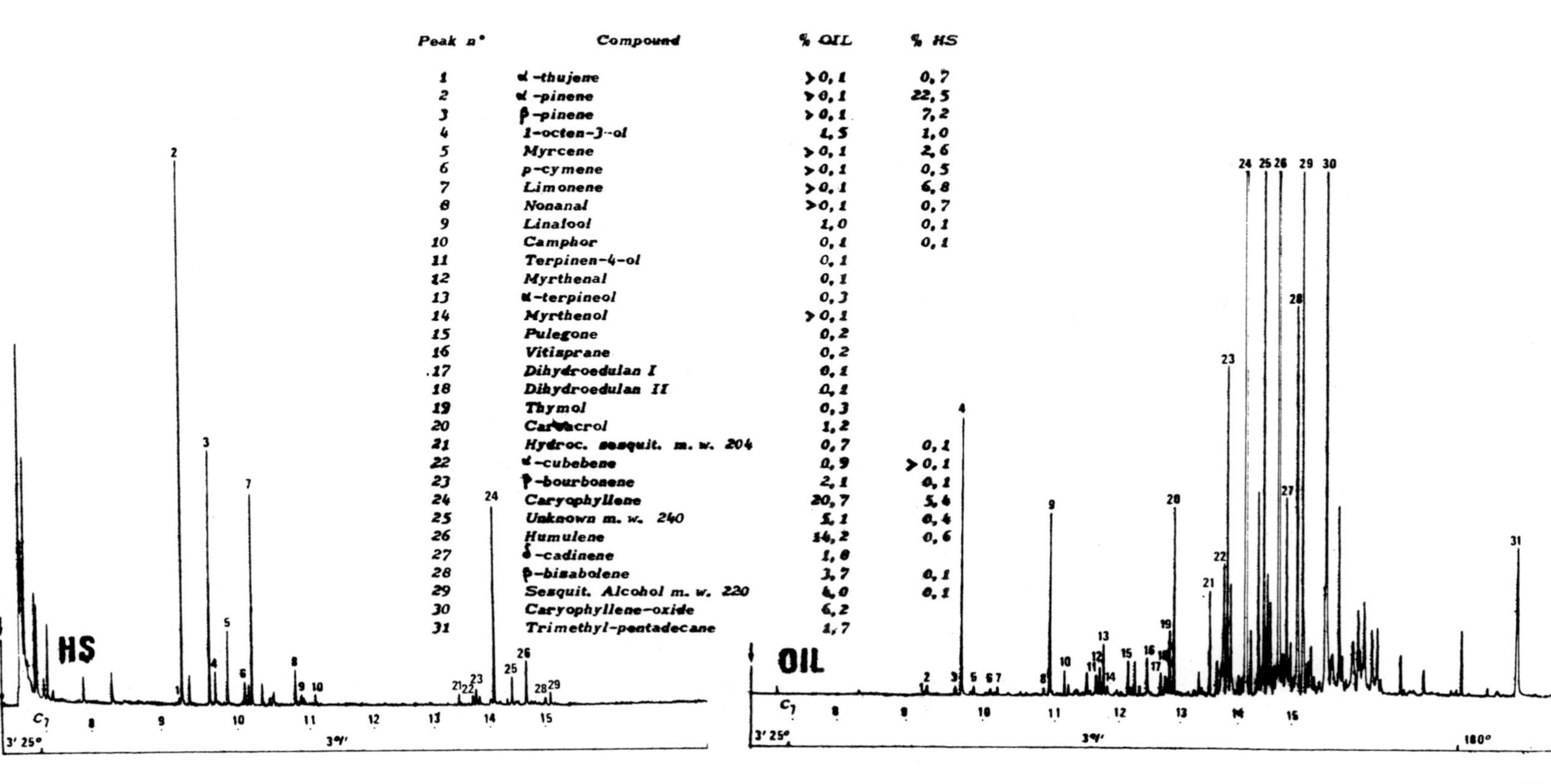

Peak n°	Compound	% OIL	% HS
1	α-thujene	>0,1	0,7
2	α-pinene	>0,1	22,5
3	β-pinene	>0,1	7,2
4	1-octen-3-ol	1,5	1,0
5	Myrcene	>0,1	2,6
6	p-cymene	>0,1	0,5
7	Limonene	>0,1	6,8
8	Nonanal	>0,1	0,7
9	Linalool	1,0	0,1
10	Camphor	0,1	0,1
11	Terpinen-4-ol	0,1	
12	Myrthenal	0,1	
13	α-terpineol	0,3	
14	Myrthenol	>0,1	
15	Pulegone	0,2	
16	Vitisprane	0,2	
17	Dihydroedulan I	0,1	
18	Dihydroedulan II	0,1	
19	Thymol	0,3	
20	Carvacrol	1,2	
21	Hydroc. sesquit. m. w. 204	0,7	0,1
22	α-cubebene	0,9	>0,1
23	β-bourbonene	2,1	0,1
24	Caryophyllene	20,7	5,4
25	Unknown m. w. 240	5,1	0,4
26	Humulene	14,2	0,6
27	δ-cadinene	1,8	
28	β-bisabolene	3,7	0,1
29	Sesquit. Alcohol m. w. 220	4,0	0,1
30	Caryophyllene-oxide	6,2	
31	Trimethyl-pentadecane	1,7	

stances appears, with a very variable distribution which depends on the type of herb. On the contrary, the corresponding HS graphs include a smaller number of compounds, particularly abundant in the initial and central zones of the chromatogram. This pattern confirms that the two series of data represent different but complementary aspects of the characteristics of a herb. Indeed, in HS the more volatile substances, on which the olfactory properties of the herb depend, are put into relief; the essential oil, on the other hand, gives the set of compounds present in the aromatic tissue of the herb. A more complete definition of the components which contribute to the aroma of the herb could be derived from the union of these values.

The most evident and remarkable characteristic of the HS profiles is the presence of compounds which are almost totally absent in the essential oils. The identification of these substances, which, to our opinion make a significant contribution to the odour and quality of the herb, is still being carried out. Their high volatility, which also prevents their recovery in essential oils, poses considerable technical problems with regards to the injector and the thermostatic system of the mass spectrometer. We think, however, that the use of special splitless and static conditioning apparatus, coupled with polar columns can contribute to the solution of the problem.

Constituents of a sesquiterpenic nature are also worthy of note, although they were only partially identified; they are in fact very abundant in essential oils but almost totally absent in head spaces. Evidently these substances, having low volatility, contribute in a minute degree to the olfactory characteristics of a herb.

More accurate studies are being made to evaluate the real olfactory contribution, in terms of the perception threshold, of the substances which characterise the odour and the essence. The results of this research will enrich the organoleptic knowledge concerning both the individual aromatic substances and the interactions existing between them; as a consequence the quality control of aromatic herbs can be achieved.

REFERENCES

1. Chialva F, Doglia G, Gabri G, Aime S, Milone L. 1976. *Rivista Italiana EPPOS* 58:522
2. Chialva F, Gabri G, Liddle PAP, Ulian F. 1981. *Rivista Italiana EPPOS* 63:286.

3. Chialva F, Liddle PAP. 1980. Paper presented at the VIII Congres International des Huiles Essentielles, Cannes (France).
4. Chialva F, Liddle PAP, Ulian F, DeSmedt P. 1980. *Rivista Italiana EPPOS* 62:297.
5. Gabri G, Chialva F. 1981. *J. High Res. Chrom. & CC* 4:215.
6. Nano GM, Sacco T, Frattini C. 1973. *Essenze Deriv. Agrum.* 43:107.
7. Nano GM, Sacco T, Frattini C. 1976. *Essenze Deriv. Agrum.* 46:171.
8. Taskinen J, Nykänen L. 1975. *Acta Chem. Scand. Sér. B* 29:425.

THE CHEMICAL COMPOSITION OF THE ESSENTIAL OIL OBTAINED FROM *SATUREJA KITAIBELII* WIERZB. ap. HEUFF.

R. PALIĆ, S. KAPOR, M.J. GAŠIĆ

1. INTRODUCTION

In chemotaxonomic studies of numerous species, infraspecific variations in the terpenoid composition are frequently noted. The general features of the monoterpenoid pattern in essential oils are considered to be primarily controlled by genetic factors while compositional variations may result from the differences in ecological conditions. Our present understanding of the relative importance of all the determinants in the formation of secondary metabolites is rather limited in view of the complexity of adequate experiments. Nevertheless, collection of data on the chemical composition of essential oils is the prerequisite and of major importance for all research related to biochemical systematics.

In this study, the composition of the monoterpenoid fraction of the essential oil obtained by steam distillation of *Satureja kitaibelii* Wierzb. ap. Heuff. was investigated; this plant was earlier considered as a subspecies of *S. montana* L. (1) but according to detailed studies by Silić it is now classified as a separate species (5). *S. kitaibelii* is distributed in the Northeastern regions of the Balkan peninsula, growing on limestone in mountain areas (ca 300-1000 m).

2. EXPERIMENTAL

Specimens were collected from representative populations of three different locations in Southeastern Yugoslavia; altitudes: ca 300 m (location I), 700 m (location II) and 1000 m (location III). The plant material was air-dried, grinded and steam distilled according to the method of Clevenger. The essential oil obtained was subjected to GC without further purification: Varian Aerograph, series 1400, FID, 25% Carbowax 20 M (3000 x 2 mm) on Chromosorb W AW, 80-100 mesh, N_2 (10 ml/min) temp. programme 2°/min. Identification of individual components was made by coinjecting authentic samples.

Margaris N, Koedam A, and Vokou D (eds.): Aromatic Plants: Basic and Applied Aspects

ISBN 90-247-2720-0. Printed in the Netherlands.

3. RESULTS AND DISCUSSION

The monoterpenoid composition of *Satureja kitaibelii* Wierzb. ap. Heuff. from locations I, II, and III is shown in Table 1 and the corresponding gas chromatograms in Figs. 1, 2, and 3, respectively. Although the number of populations examined so far is insufficient for definite conclusions, the monoterpenoid composition, i.e. the concentration of major diagnostic components, indicate the existence of two distinct chemotypes: geraniol-

Table 1. Monoterpenoids in *S. kitaibelii* Wierzb. ap. Heuff.

Components %, Peak No (GC)	Location I (Fig. 1)	Location II (Fig. 2)	Location III (Fig. 3)
α-pinene	0.60 (1)	0.91 (1)	1.01 (1)
camphene	0.31 (3)	0.36 (4)	0.55 (2)
β-pinene	-	0.05 (6)	0.05 (3)
Δ^3-carene	0.07 (5)	-	0.09 (4)
myrcene	0.37 (6)	0.26 (8)	0.08 (5)
p-cineole	0.20 (8)	-	0.54 (7)
limonene	2.85 (9)	5.32 (11)	7.28 (8)
1.8-cineole	0.58 (10)	0.71 (12)	1.89 (9)
γ-terpinene	1.15 (11)	0.78 (14)	2.59 (10)
p-cymene	1.45 (12)	1.31 (15)	23.11 (11)
menthone	0.25 (19)	0.82 (22)	0.28 (18)
linalool	5.06 (20)	29.89 (23)	12.69 (19)
linalyl acetate	-	-	0.26 (20)
camphor	0.46 (21)	-	-
terpinen-4-ol	10.79 (22)	5.70 (25)	5.60 (22)
α-terpineol	3.53 (24)	5.30 (28)	8.59 (26)
borneol	1.60 (25)	-	-
geranyl acetate	2.18 (26)	30.05 (29)	1.05 (27)
carvone	-	-	0.67 (28)
nerol	1.22 (27)	0.46 (30)	0.60 (29)
geraniol	58.69 (28)	13.69 (31)	0.60 (31)
thymol	2.52 (32)	0.91 (33)	19.55 (36)
carvacrol	-	-	2.19 (37)

linalool-terpinen-4-ol chemotype, locations I and II, and *p*-cymene-thymol chemotype for location III. As can be seen, the concentration of phenolic components, particularly of carvacrol, is relatively low (even in location III) and thus in contrast with earlier findings regarding terpenoid composition of various Mediterranean plants of the genus *Satureja* (4).

Investigation of essential oil yields obtained from plants at different developmental stages show continous decrease, the yields being high at the stage of actively growing shoots and lowest after the blooming period (Fig. 4); the decrease of oil yields is much faster than that found for *S. hortensis* L. and *S. montana* L. (2,3) . Interestingly, plants from the three ecologically different locations yield the same quantity of oil in the blooming period. It should also be noted that the concentrations of the main components, geraniol (and geranyl acetate) and linalool, change substantially over the vegetation period in opposite directions (Figs. 5, 6).

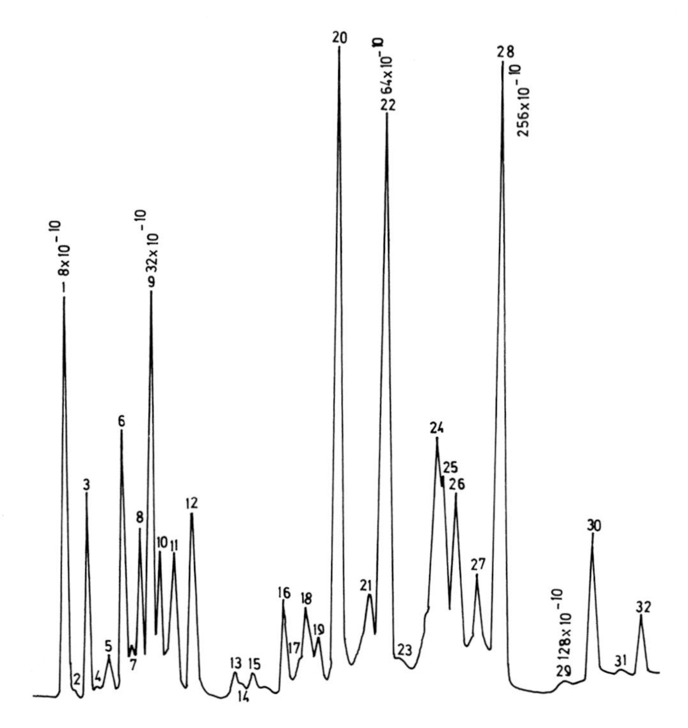

FIGURE 1. GC of the essential oil from location No I.

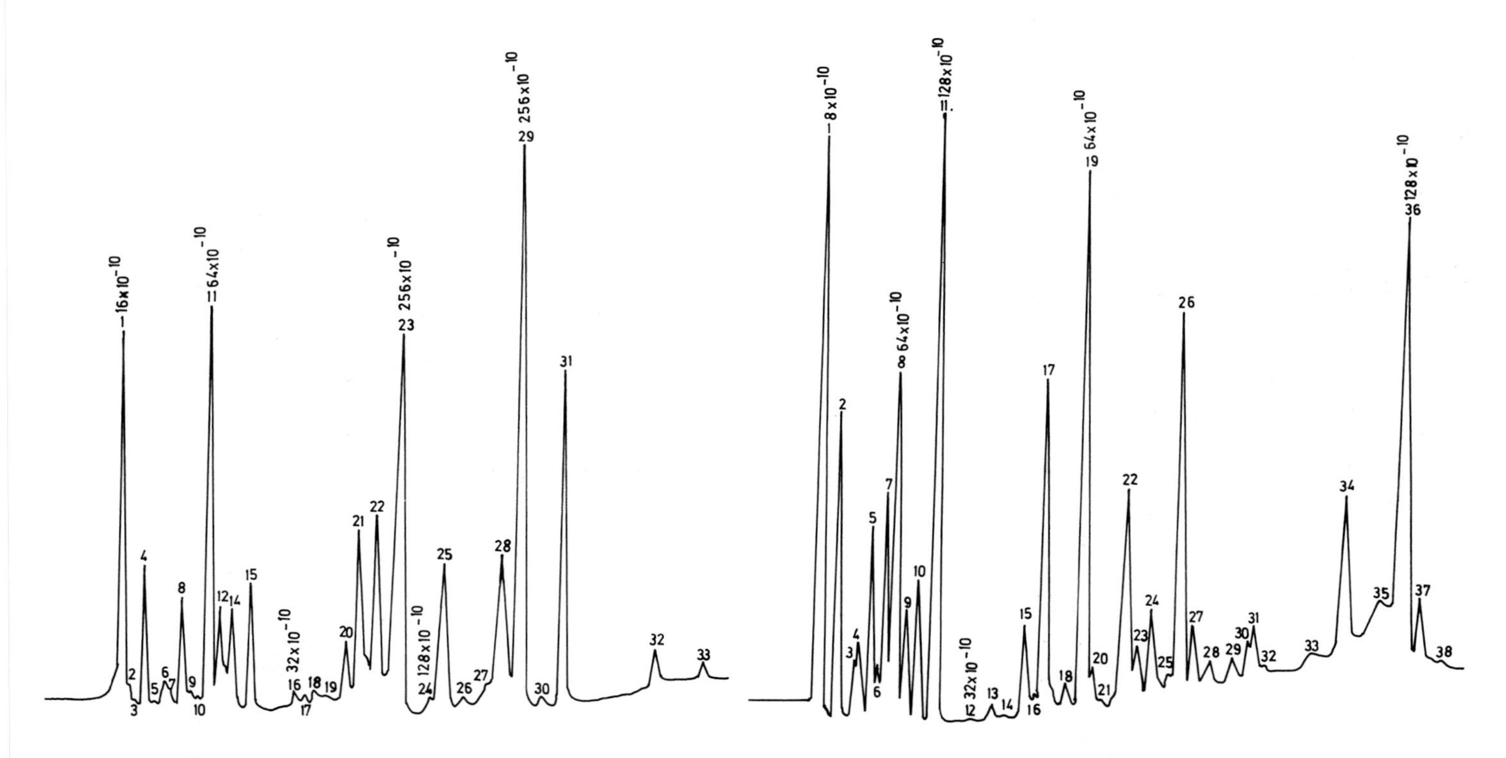

FIGURE 2. GC of the essential oil from location No II. FIGURE 3. GC of the essential oil from location No III.

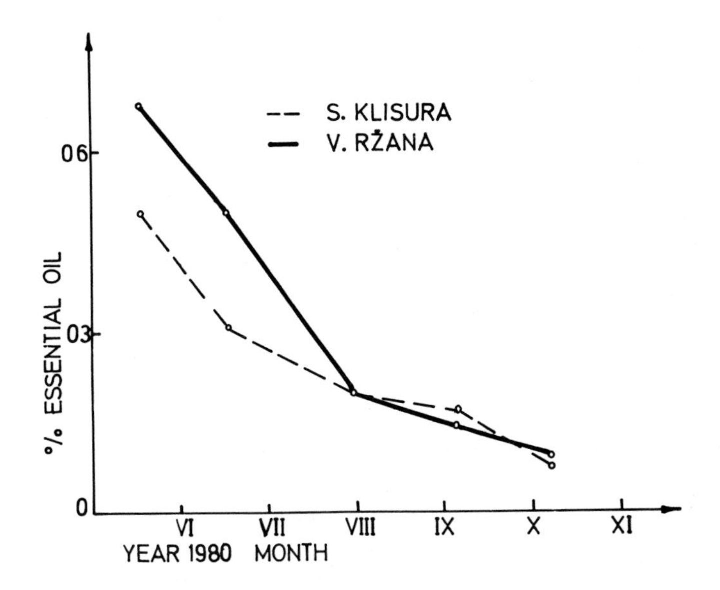

FIGURE 4. Seasonal variations in essential oil yields. (S. Klisura and V. Ržana are two sites in Yugoslavia where plants were collected).

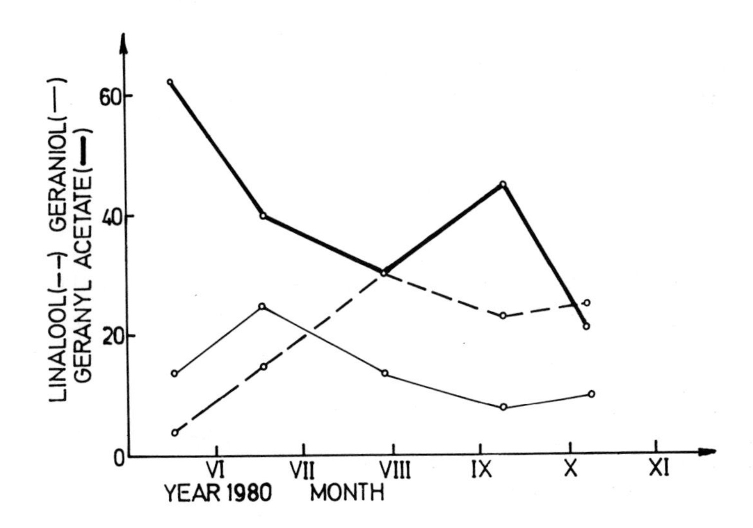

FIGURE 5. Seasonal variations in yields of geraniol, geranyl acetate and linalool.

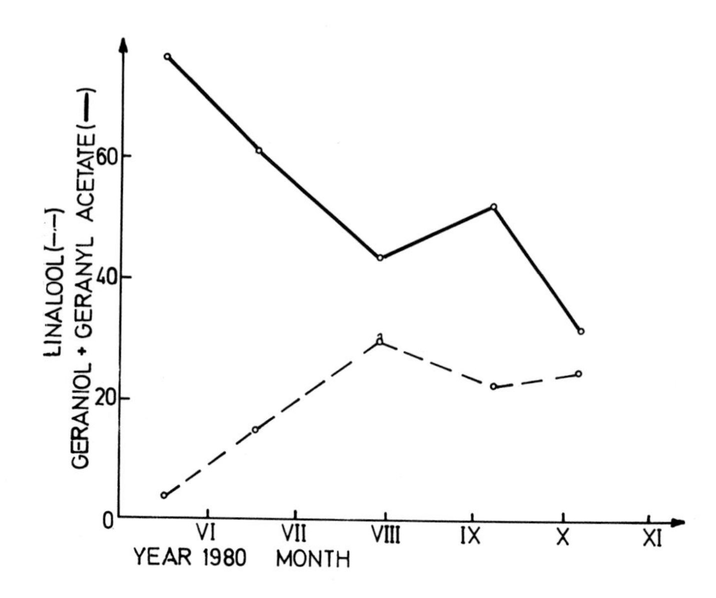

FIGURE 6. Seasonal variations in yields of geraniol and geranyl acetate vs. linalool.

In a separate experiment, GC analyses of the hydrocarbon fraction obtained by pentane extraction of the plant material were also carried out: C_{17}-C_{35} hydrocarbons were identified, with an odd/even ratio of 7:1. These results, in relation to hydrocarbon fraction composition obtained from other plants of genus *Satureja* will be published elsewhere.

The obtained chemical evidence offers additional support for the classification of *S. kitaibelii* Wierzb. ap. Heuff. as a separate species.

REFERENCES

1. Ball PW, Gentiffe MF, Tutin TG. 1972. Flora Europaea Vol. III, p. 164. Cambridge, University Press.
2. Nikolajev AG, Andrejeva Y. 1956. *Uchenye Zapiski Kishinev Univ.* 23:69.
3. Palić R, Kapor S, Gašić MJ. 1980. II Yugoslav Symposium in Organic Chemistry.
4. Thieme H, Thi Tam N. 1972. *Pharmazie* 27:255 and references cited therein.
5. Silić Č. 1979. Monographie der Gattungen *Satureja* L., *Calamintha* Miller *Micromeria* Bentham, *Acinos* Miller und *Clinopodium* L. In: Flora Jugoslaviens. Sarajevo, Svjetlost.

THE ESSENTIAL OIL FROM *THYMUS PRAECOX* SSP. *ARCTICUS*

EL. STAHL

1. INTRODUCTION

Thymus praecox ssp. *arcticus* (E. Durand) Jalas is one of the most widely distributed plants in Iceland. It is common in the lowland as well as in the central highland up to an altitude of 1000 m above sea level. It is growing in virtually all kinds of localities, i.e. on dry soil like lava fields, sandy and gravelly soil, in heaths and near hot springs. The decoction is used as a remedy against cold, cough, pulmonary and heart diseases, epilepsy, hiccup, insomnia etc. (1).

2. PROCEDURE

2.1. Material and Methods

The plant material was collected at different places in the south and north of Iceland. The essential oil was obtained from fresh herb by steam distillation (yield 0.13%). After prefractionation of the obtained oil by column chromatography over silica gel (2) the received fractions were analysed by capillary GLC and combined GLC-MS.

2.1.1. Gas chromatography. Perkin Elmer F 22 with FID and integrator Spectra Physics System I. 22 m glass capillary WG 11, nitrogen 1.2 ml min^{-1}, inj. temp. 180°C, det. temp. 200°C. Column temperature was programmed from 80°C to 200°C at 2°C min^{-1}. The quantitative results were calculated as peak area percentages without correction factor.

2.1.2. Mass spectrometry. Hitachi - Perkin Elmer RMU D6, 70 eV.

3. RESULTS

The composition of the essential oil of *Thymus praecox* ssp. *arcticus* is shown in Fig. 1 and the components are listed in Table 1. Unlike in other *Thymus* species, neither thymol nor carvacrol could be found in the essential oil of this species. On the other hand, the essential oil contains three higher boiling compounds (peak 90, 93 and 94), whose chromatographic data and MS data prove to be nerolidol (Fig. 2), elemol ? (Fig. 3)

Margaris N, Koedam A, and Vokou D (eds.): Aromatic Plants: Basic and Applied Aspects

ISBN 90-247-2720-0.
Printed in the Netherlands.

Table 1. Components of the herb oil of *Thymus praecox* ssp. *arcticus*.

No	Compounds	Fraction*	Percentages†	Identification
3	α-Pinene	1	0.04	GC
5	Unknown	1	0.02	
7	β-Pinene	1	0.02	GC
8	Sabinene	1	0.02	GC
12	Myrcene	1	2.26	GC
15	Limonene	1	0.46	GC
17	1.8-Cineole	4	0.50	GC,MS
18	*cis*-Ocimene	1	1.00	GC
20	*trans*-Ocimene	1	2.10	GC
24	Terpinolene	1	1.00	GC
26	1-Octene-3-yl acetate	3	1.40	GC,MS
30	1-Octene-3-ol	5	0.79	GC,MS
31	Unknown	5	0.51	
37	Camphor	4	0.53	GC,MS
38	Linalool	4/5	27.42	GC,MS
39	Linalyl acetate	3	25.37	GC,MS
47	Caryophyllene	1	2.16	GC,MS
61	α-Terpineol	5	7.38	GC,MS
63	Germacrene-D	1	0.95	GC,MS
65	Neryl acetate	3	2.92	GC,MS
66	β-Bisabolene	1		GC,MS
75	Geranyl acetate	3	3.84	GC,MS
76	γ-Cadinene	1	0.37	GC,MS
77	Unknown (MW 204)	1	1.16	MS
79	Nerol	5	4.04	GC,MS
82	Geraniol	5	1.43	GC,MS
90	Nerolidol	4	1.20	GC,MS
92	Unknown (MW 220)	4	0.20	MS
93	Elemol (?)	5	2.49	MS
94	Unknown (MW 220)	4	4.59	MS
		Total	96.42	

* see reference (2)

† *Thymus* type A

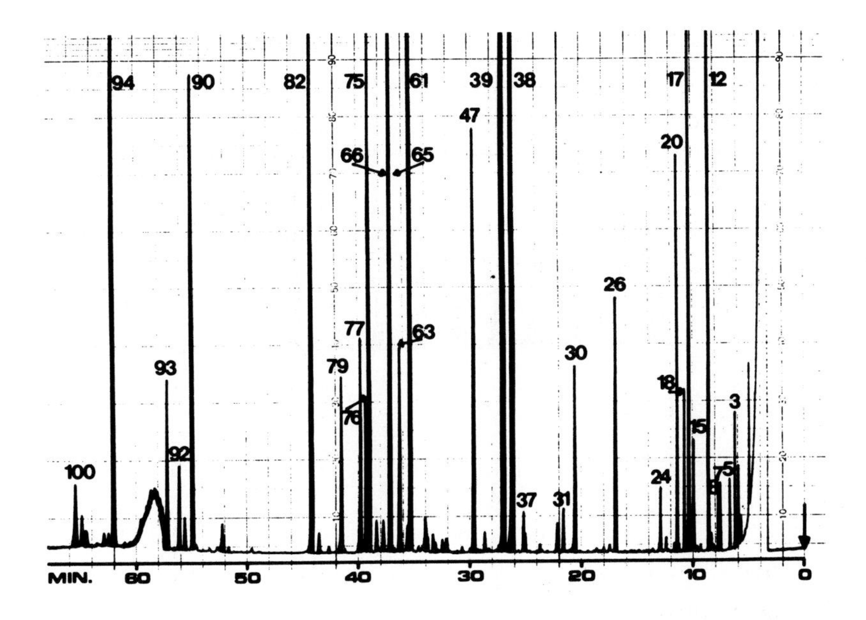

FIGURE 1. Gas chromatogram of the herb oil of *Thymus praecox* ssp. *arcticus* (E. Durand) Jalas.

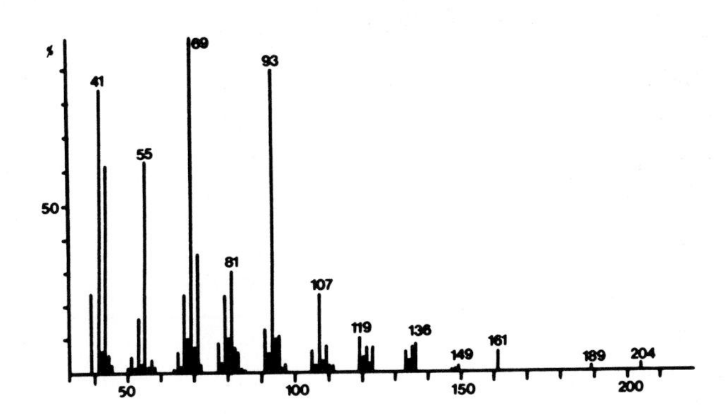

FIGURE 2. MS of compound No. 90. Nerolidol

and one till now unknown sesquiterpene alcohol (Fig. 4).

The quantitative variation of these three compounds in the essential oil received from plants of twenty eight places of Iceland show that there exist at least six types of plants (Table 2). A correlation between plant type and locality, climatic or edaphic factors could not yet be established.

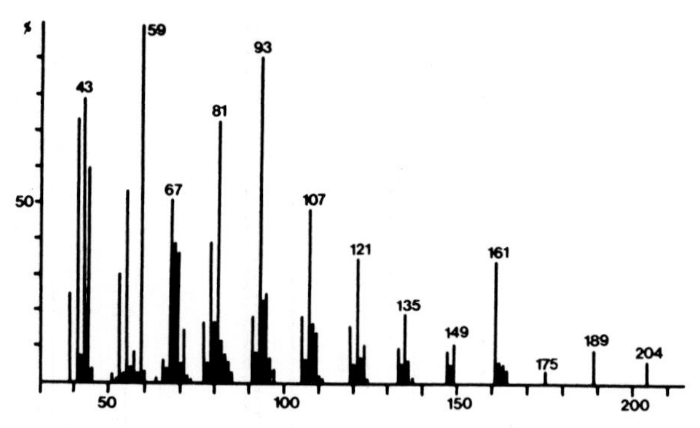

FIGURE 3. MS of compound No. 93. Elemol (?).

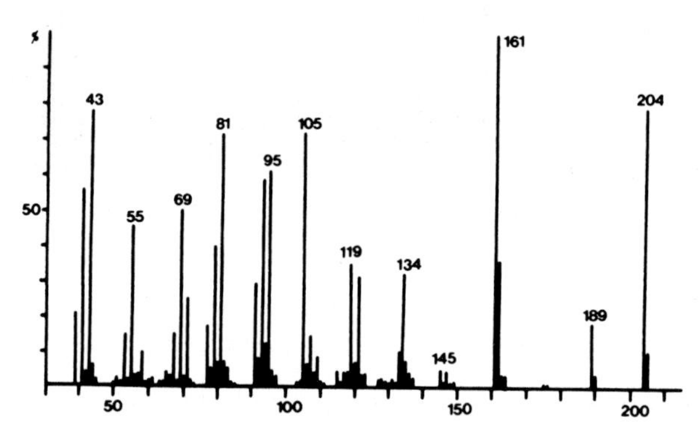

FIGURE 4. MS of compound No. 94. Unknown.

Table 2. Characterization of the chemical types of *Thymus praecox* ssp. *arcticus* (E. Durand) Jalas.

Type	Compound 90 Nerolidol	Compound 93 Elemol (?)	Compound 94 Unknown	Frequency
A	+	+	+	11
B	-	++	++	9
C	++	+	+	3
D	++	-	-	3
E	+	+	++	1
F	-	-	-	1
			Total	28

+ = 0.1 - 5.9% ; ++ = 6.0 - 25%

REFERENCES

1. Jónsson BL. 1977. Islenskar Laekninga- og Drykkjarjurtir. Reykjavik, Önnur prentum.
2. Kubeczka K-H. 1973. *Chromatographia* 6:106.

ATRACTYLODES LANCEA DC. (COMPOSITAE). CONTRIBUTION ON THE CONSTITUENTS OF THE ESSENTIAL OIL

K. BRUNS, H. DOLHAINE, U. WEBER.

1. INTRODUCTION

A. lancea DC., identical with *A. ovata* Thunb. is indigenous in China and Japan. The essential oil is obtained by water-steam distillation of the roots and stalks of the plant; the concrete oil by extraction with volatile solvents.

Because of the odour properties of the oil, the plant has been cultivated; both the essential and the concrete oils are commercially available. Due to its characteristic blossom-odour the oil is used in perfumery and the soap industry since a long time.

Up to now the essential oil has been the object of several investigations which led to the identification of the following constituents:

Hydrocarbons: *p*-cymene, α-curcumene, β-selinene (1, 8)
aromadendrene (11, 16, 17)

Alcohols : atractylol (mixture of β-eudesmol/hinesol) (1, 2, 3, 4, 6, 7, 8, 9, 10, 13, 15, 16, 19, 21, 23, 25, 26).

Furanoids : furfurol (3, 7)
atractylon (1, 3, 5, 8, 10, 13, 16, 18, 20)
3β-hydroxyatractylon, 3β-acetoxyatractylon (12, 14),
atractylodin (7, 24).

With the exception of furfurol and atractylodin, all constituents belong to the monoterpene and sesquiterpene series.

2. EXPERIMENTAL

The investigations were carried out with the following instrumental equipment:

GLC: Finnigan 9610
50 m Glasscapillar WG 11
80°-3°-220°C
Split 1:100

Margaris N, Koedam A, and Vokou D (eds.): Aromatic Plants: Basic and Applied Aspects
ISBN 90-247-2720-0.
Printed in the Netherlands.

Mass-Spectrometer: Finnigan 4000
Ionisation-energy 70 eV
Scan 35-350
Scan-Time 0.5 sec.

Data-System : Incos 2300

The experimental material was a commercial oil of Chinese origin with the following analytical data:

d_4^{20}	0.9339	n_D^{20}	1.5032
α_{589}^{22}	$+34.55^o$		
Acid value	3.4		
Saponification value	10.5		
Ester value	7		

The material was separated into different fractions with respect to the functional character of the constituents. The following fractions were obtained:

	%
Acids	0.12
N-Bases	0.03
Phenolics	0.3
Ester	0.8
Carbonyl Compounds	0.3
Monoterpenes	12.5
prim./sec. Alcohols	4.0
tert. Alcohols	42.0
Sesquiterpenes + probably Lactones	28.0
work-up loss	11.95

The identification of the constituents in the different fractions resulted from GLC/mass-spectroscopic comparison with authentic samples, in some cases from comparison with the mass-spectra from the current literature (indicated by *).

3. RESULTS

The following new constituents could be identified:

Monoterpenes	Sesquiterpenes	Terpenoids
α-pinene	β-elemene	a) Alcohols
camphene	caryophyllene	borneol
β-pinene	muurolene*	citronellol
sabinene	valencene	nerol
3-carene	δ-cadinene*	geraniol
myrcene		α-terpineol
α-phellandrene	Phenolics	*cis*(*trans*)-sabinol*
α-terpinene	*o*-cresol	b) Ketones
limonene	*p*-cresol	carvotanaceton
cis-β-ocimene	thymol	valeranon*
trans-β-ocimene	carvacrol	(= jatamanson)
terpinolene		c) Esters
		(as acids from the saponifiable fraction)
		cis-geranium acid
		trans-geranium acid

non-Terpenoids

a) (as acids from the saponifiable fraction)
   isovalerianic acid
   capronic acid

b) n-dodecanol

In addition to the above listed new constituents we could identify the following compounds, already described in the literature:

p-cymene	elemol	
β-selinene	β-eudesmol hinesol	= atractylol

For the quantities of the main-constituents, β-eudesmol/hinesol, we found a ratio of 2.6:1 (literature: 1.5:1)

ACKNOWLEDGEMENTS

We are grateful to Mr. A. Zibula for help with the mass spectra and to Dr. Lamparsky for samples and mass spectra of *cis*-/*trans*-geranium-acid-methylester.

REFERENCES

1. Chow WZ, Motl O, Sŏrm F. 1962. *Coll. Czech. Chem. Comm.* 27:1914.
2. Gadamer J, Amenomiya T. 1903. *Arch. Pharm.* 241:22.
3. Gildemeister E, Hoffmann Fr, 1961. Die Ätherischen Öle. Bd. VII b, p. 760. Berlin, Akademie Verlag.

4. ibid. 1962. Bd. III b, pp. 270, 294.
5. ibid. 1963. Bd. III c, p. 464.
6. Karrer W. 1958. Konstitution und Vorkommen der organischen Pflanzenstoffe, p. 759. Basel-Stuttgart, Birkhäuser Verlag.
7. ibid. 1977. Ergänzungsband 1, pp. 187, 203, 205, 526, 808.
8. ibid. 1981. Ergänzungsband 2, Teil 1, pp. 23, 24, 424, 426.
9. Lafontaine J, Morgain M, Sergent-Guay M, Ruest L. 1980. *Can. J. Chem.* 58:2460.
10. Miltitzer Berichte. 1967/68, 12.
11. ibid. 1972, 12.
12. ibid. 1977, 182.
13. Motl O. Chow WZ, Sŏrm F. 1961. *Chem. Ind.* 207.
14. Nisikawa Y, Watanabe Y, Seto T, Yasuda I. 1976. *Yakugaku Zasshi* 96:1089 (*C.A.* 1976. 85:192927g).
15. Ruzicka L, Koolhaas DR, Wind AH. 1931. *Helv. Chim. Acta* 14:1178.
16. Studennikova LD, Chaleckij AM. 1966. *Apecnoe Delo* 15:27 (*Biol. Abstr.* 48:108658).
17. Studennikova LD, Chaleckij AM. 1971. *Rast. Resur.* 7:396 (*C.A.* 1971. 75:143900c).
18. Takagi S, Hongo G. 1924. *Yakugaku Zasshi* 44:539.
19. Takahashi S, Hikino H, Sasaki Y. 1959. *Yakugaku Zasshi* 79:541 (*C.A.* 1959. 53:15479h)
20. Takahashi S, Hikino H, Sasaki Y. 1959. *Yakugaku Zasshi* 79:544 (*C.A.* 1959. 53:15479i).
21. Ueno K. 1892. *Yakugaku Zasshi* 129:1074.
22. Wells FW. 1959. *Soap Perfum. Cosmetics* 32:261.
23. Yosioka I, Hikino H, Sasaki Y. 1959. *Chem. Pharm. Bull.* 7:817.
24. Yosioka I, Hikino H, Sasaki Y. 1960. *Chem. Pharm. Bull.* 8:949, 952, 957.
25. Yosioka I, Sasaki Y, Hikino H. 1961. *Chem. Pharm. Bull.* 9:84.
26. Yosioka I, Takahashi S, Hikino H, Sasaki Y. 1959. *Chem. Pharm. Bull.* 7:319.

CHEMOTAXONOMY OF THE GREEK SPECIES OF *SIDERITIS*
I. Components of the volatile fraction of *Sideritis raeseri* ssp. *raeseri*

V.P. PAPAGEORGIOU, S. KOKKINI, N. ARGYRIADOU

1. INTRODUCTION

As it has been reported (3), we have started studying the Greek species of the genus *Sideritis*, section Empedoclia, from a biosystematic point of view. Furthermore we are studying the chemical differences or similarities between these species, on the basis of the components of the volatile fraction, used for the first time for such an investigation.

Venturella et al. investigated the diterpenes of some species of *Sideritis* and found that there is a chemical distinction between *Sideritis euboea* Heldr., *S. raeseri* Boiss. & Heldr., *S. syriaca* L. and *S. clandestina* (Bory & Chaub.) Hayek (4).

In the case of *S. raeseri* we distinguished three subspecies: *S. raeseri* ssp. *raeseri*, *S. raeseri* ssp. *florida* (Boiss. & Heldr.) Pap. & Kok., and *S. raeseri* ssp. *attica* (Heldr.) Pap. & Kok.

The morphological differences of these subspecies are shown in Table 1.

Table 1. Morphological differences between three subspecies of *Sideritis raeseri*.

Character	ssp. *raeseri*	ssp. *florida*	ssp. *attica*
Height of plant	30-45 cm	30-45 cm	15-35 cm
Density of inflorescence	distant	± dense	distant
Bracts	18-20x15-17mm, equal or shorter than flowers with acumen 5-7mm.	20-24x19-21mm, equal or exceeding flowers with acumen 7-9mm.	6-8x8-10mm, shorter than flowers with acumen 1-2mm.

2. MATERIAL AND METHODS

The aerial parts of the plant collected from Mt Sinniatsikon (NW GREECE) were subjected to steam distillation and yielded 0.1% of essential oil. Analysis of the oil was carried out by using a computerised gas chromato-

Margaris N, Koedam A, and Vokou D (eds.): Aromatic Plants: Basic and Applied Aspects

ISBN 90-247-2720-0. Printed in the Netherlands.

graphic spectrometric system Hewlett-Packard 5989 A (data system).

Gas chromatographic separations were performed on a 12 ft glass column packed with 3% of Carbowax 20M (Chromosorb W AW DMCS, 80-100 mesh). The injector temperature was 100°C, the FID detector was heated to 300°C and the column temperature was programmed between 80-230°C at the rate of 2°C min^{-1}. Helium was the carrier gas flowing at 35 ml min^{-1}. Mass spectra were taken utilizing an ionizing voltage of 70 eV.

Identification of the individual components was accomplished with the aid of various computer interpretative techniques, as well as by individual interpretation of some spectra.

3. RESULTS AND DISCUSSION

By the chemical analysis of the essential oil of *Sideritis raeseri* ssp. *raeseri* (chromosome number 2n = 32) 60 constituents were isolated, of which 50 could be identified. The gas chromatogram is shown in Fig. 1, the chemical composition of the essential oil in Table 2 and the chemical structure of the identified compounds in Fig. 2.

It is worth noticing the high concentration of naphthalene (~22%) not only in the case of *S. raeseri* ssp. *raeseri* but furthermore, in the essential oils of other species of the genus *Sideritis*, such as *S. syriaca* L. and *S. euboea* Heldr. (2). Nevertheless, it is our opinion that the formation of naphthalene may be attributed to an artifact product which is formed during steam distillation (1).

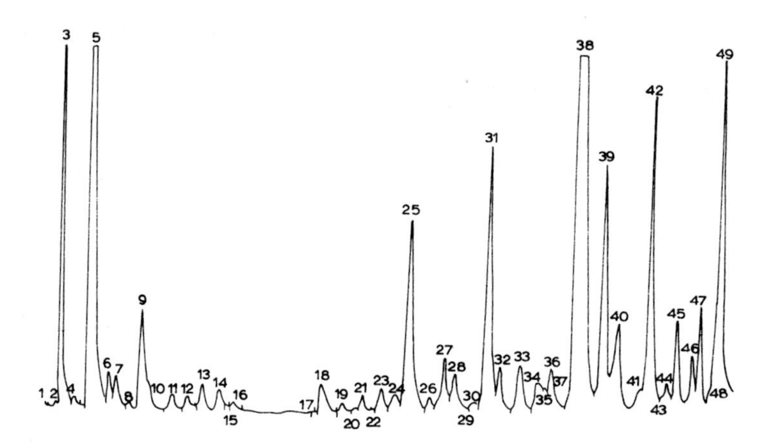

FIGURE 1. Gas chromatogram of the volatile oil of *Sideritis raeseri* ssp. *raeseri*.

Table 2. Chemical composition of essential oil of the whole plant *Sideritis raeseri* ssp. *raeseri*.

No of peak	R_t (min.)	M^+	m/e	Components	%
1	2.47	32	15,31,32 100%⟶	Methanol	0.016
2	3.30	—	57,43,71,85 100%⟶	Hydrocarbon	0.026
3	3.76	136	93,91,92,77,79,121,136 100%⟶	α-Pinene	2.814
4	4.72	136	93,67,121 100%⟶	Camphene	0.056
5	5.97	136	93,91,69,79,77,121,136 100%⟶	β-Pinene	11.187
6	7.39	136	93,91,92,77,136 100%⟶	α-Phellandrene	0.263
		136	93,69,41,91,79,77 100%⟶	Myrcene	
7	7.92	130	74,87 100%⟶	Methyl caproate	0.259
		158	57,85,41,56,103 100%⟶	Butyl valerate	
		136	93,121,91,77,79,136 100%⟶	α-Terpinene	
8	8.82	172	41,43,71,70,85,57 100%⟶	Isoamyl isovalerate	0.048
9	9.85	136	68,67,93,136,121,39,41 100%⟶	Limonene	1.130
10	10.25	136	93,136,77,91 100%⟶	β-Phellandrene	traces

Table 2 (Continued)

No of peak	R_t (min.)	M^+	m/e	Components	%
11	12.10	136	93,91,92,79,77,121,136 100%⟶	Ocimene	0.165
		138	81,82,53,138,27,39,29 100%⟶	2-Pentyl furan	
12	13.25	136	93,91,79,77,80,92,121,136 100%⟶	γ-Terpinene	0.105
13	14.35	134	119,134,91,105 100%⟶	*p*-Cymene	0.205
14	15.58	172	70,57,85,43,41,55,71,103 100%⟶	Isoamyl valerate	0.178
		144	74,87 100%⟶	Methyl enanthate	
15	16.03	136	93,121,136,79,77 100%⟶	Terpinolene	0.017
		—	84,56,57,69,41,43 100%⟶	Aldehyde (C_8)	
16	16.71	172	85,70,57,43,41,71,55,103 100%⟶	Amyl valerate	0.052
17	22.51	130	55,59,83,101 100%⟶	3-Octanol	0.003
18	23.01	—	57,41,43,44,56,55,70,69,82,98 100%⟶	Aldehyde (C_9)	0.319
19	24.61	—	—	Unidentified	0.071
20	25.61	—	—	Unidentified	0.026

Table 2 (Continued)

No of peak	R_t (min.)	M^+	m/e	Components	%
21	26.07	128	57,43,72,41,85,55 100%⟶	1-Octen-3-ol	0.120
		186	85,103,84,69 100%⟶	Hexyl valerate	
22	26.87	147	146,148 100%⟶	Dichlorobenzene	traces
23	27.41	184	82,67,57,41,85,55 100%⟶	3-Hexenyl isovalerate	0.209
		154	69,112,41,55,70 100%⟶	Menthone	
		204	161,119,105,204 100%⟶	β-Copaene	
24	28.44	184	57,82,67,41,85,55 100%⟶	3-Hexenyl valerate	0.206
25	29.41	204	161,119,105,204,41,93 100%⟶	α-Copaene	2.230
26	30.91	204	81,80,123,79,161,41,77,91 100%⟶	Bourbonene	0.125
27	32.02	154	71,43,41,93,55,69,80,67 100%⟶	Linalool	0.509
28	32.77	—	56,55,69,70,41,43,83,84 100%⟶	Alcohol (C_8)	0.402
29	33.65	196	95,93,136,121,43,108,110 100%⟶	Bornyl acetate	traces
30	34.12	154	71,43,93,111,69,105,107 100%⟶	Terpinen-4-ol	0.012

Table 2 (Continued)

No of peak	R_t (min.)	M^+	m/e	Components	%
31	35.23	204	93,133,91,79,69,105,107 100%→	Caryophyllene	2.560
32	36.09	150	79,107,77,91,106,105,108,135 100%→	Myrtenal	0.376
33	37.55	152	92,55,70,91,83,41,69,81,109 100%→	*trans*-Perillyl alcohol	0.451
		—	56,55,70,69,41 100%→	Alcohol (C_9)	
34	38.91	182	82,83,67,55,41,39 100%→	*cis*-3-Hexenyl tiglate	0.280
35	39.36	204	93,41,80,121,204,53,91,56 100%→	α-Humulene	0.169
36	39.89	154	59,93,121,136,81,67 100%→	Terpineol	0.470
37	40.54	204	161,105,41,91,81,119,79,93 100%→	Germacrene-D	traces
38	41.88	128	128,127,129,51,64,126,63,102 100%→	Naphthalene	21.798
39	41.92	204	41,69,93,204,94,79,55,109 100%→	Bisabolene	traces
40	43.69	204	161,105,119,41,91,134,204,77 100%→	δ-Cadinene	2.324
41	44.77	204	134,161,119,204,105 100%→	Cadinene isomer	1.263

Table 2 (Continued)

No of peak	R_t (min.)	M^+	m/e	Components	%
42	46.88	—	—	Unidentified	0.259
43	47.07	202	159,128,129,160,202,115,131,144 100%⟶	Calamene	2.842
44	48.04	—	—	Unidentified	0.083
45	48.34	191	91,108,57,85,192 100%⟶	Benzyl valerate	0.242
46	49.01	—	—	Unidentified	0.792
47	50.12	200	157,142,141,200 100%⟶	Calacorene	0.495
48	50.69	—	—	Unidentified	0.763
49	51.46	—	—	Phenyl ethyl valerate	traces
50	52.13	—	—	Sesquiterpene ester	4.290

FIGURE 2. Chemical structure of the compounds of the essential oil of *Sideritis raeseri* ssp. *raeseri*.

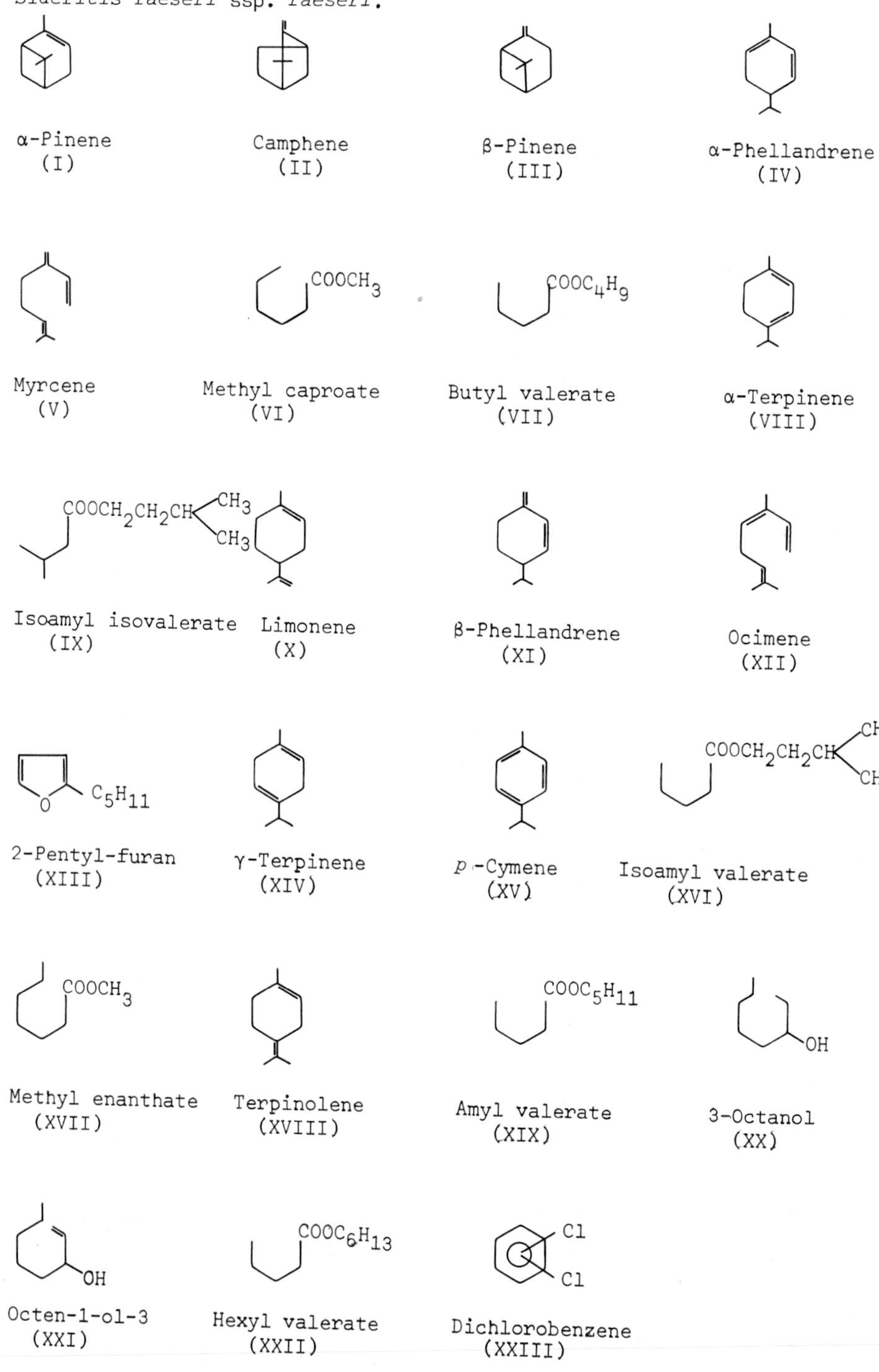

FIGURE 2. (Continued)

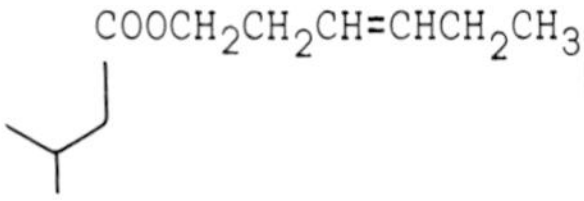

COOCH2CH2CH=CHCH2CH3

3-Hexenyl isovalerate
(XXIV)

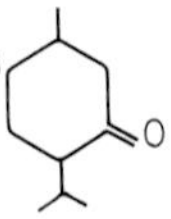

O

Menthone
(XXV)

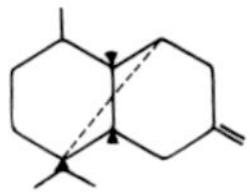

β-Copaene
(XXVI)

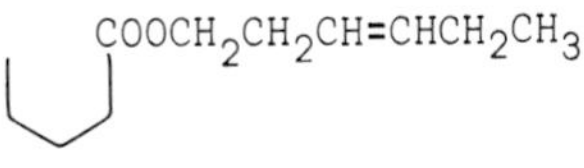

COOCH2CH2CH=CHCH2CH3

3-Hexenyl valerate
(XXVII)

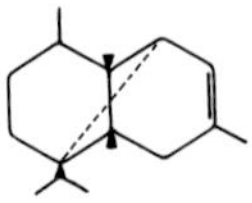

α-Copaene
(XXVIII)

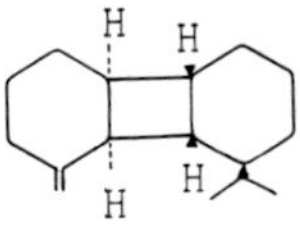

H
H
H
H

Bourbonene
(XXIX)

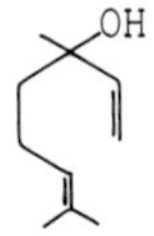

OH

Linalool
(XXX)

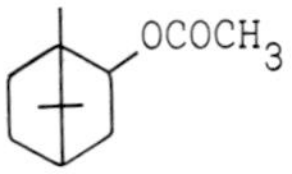

OCOCH3

Bornyl acetate
(XXXI)

OH

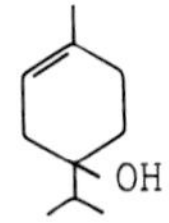

Terpinen-4-ol
(XXXII)

H
H

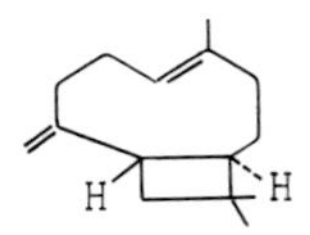

Caryophyllene
(XXXIII)

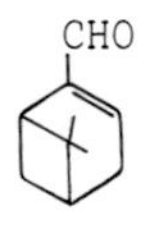

CHO

Myrtenal
(XXXIV)

CH2OH

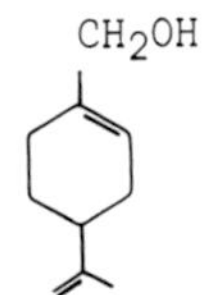

trans-Perillyl alcochol
(XXXV)

COOCH2CH2CH=
CHCH2CH3

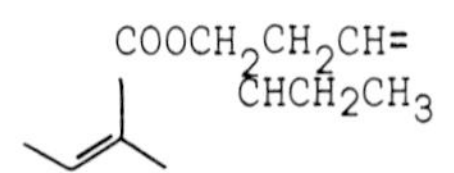

cis-3-Hexenenyltiglate
(XXXVI)

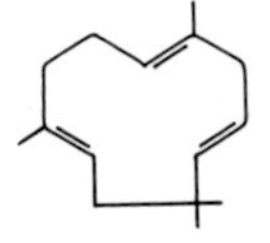

α-Humulene
(XXXVII)

OH

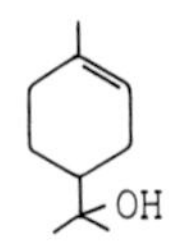

Terpineol
(XXXVII)

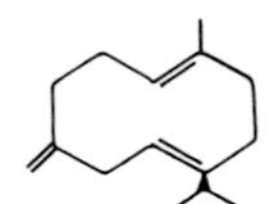

Germacrene-D
(XXXIX)

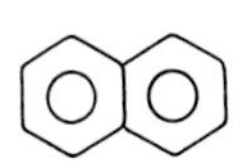

Naphthalene
(XXXX)

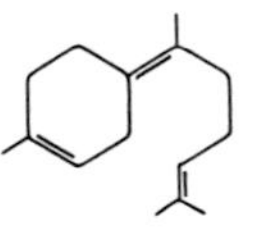

Bisabolene
(XXXXI)

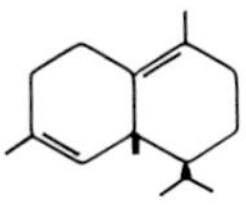

δ-Cadinene
(XXXXII)

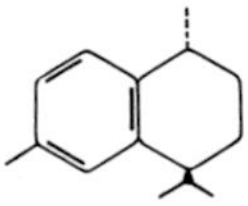

Calamene
(XXXXIII)

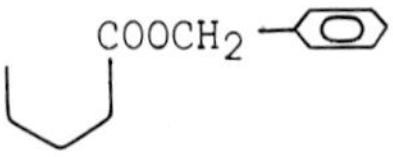

COOCH2

Benzyl valerate
(XXXXIV)

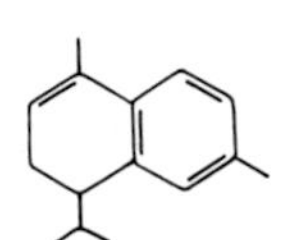

Calacorene
(XXXXV)

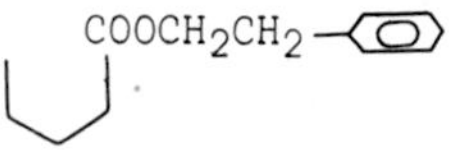

COOCH2CH2

Phenyl ethyl valerate
(XXXXVI)

REFERENCES

1. Heyns K, Stute R, Paulsen H. 1966. *Carbohydrate* 2:132.
2. Papageorgiou VP, Argyriadou N, Kokkini S. 1981. Unpublished data.
3. Papanikolaou K, Kokkini S. 1982. This volume, 131.
4. Venturella P, Bellino A. 1977. *Fitoterapia* 48:3.

THE TERPENES OF THE ESSENTIAL OIL OF MYRRH

C.H. BRIESKORN, P. NOBLE

1. INTRODUCTION

Compared to its famous applications in antiquity, nowadays myrrh has only little importance. Due to its essential oil, myrrh enters as an ingredient in spices, skin ointments and perfumes, and its tincture is used against inflammations of the oropharynx. Myrrh contains up to 10% of essential oil, responsible not only for its odour but also for its bitter taste.

2. PROCEDURE

Myrrh was exctracted by n-hexane. The components of this extract were separated by TLC, GLC, preparative GLC and GLC with middle pressure.

In the literature, components such as α-pinene, d,l-limonene, eugenol, m-cresol, curcumin aldehyde and cinnamic aldehyde are quoted. However, we could detect only sesquiterpenes (3) in a sample of myrrh, according to the prescription of the German Pharmacopoeia DAB 8; none of the above compounds was found.

The isolated sesquiterpenes were identified by IR-, ^{1}H-NMR-, high frequency ^{1}H-NMR-, ^{13}C-NMR-, UV-, mass-spectrometry, co-injection of authentic samples and synthesis of derivatives.

The sesquiterpene hydrocarbons found in myrrh were: δ-elemene, β-elemene, α-copaene, β-bourbonene, germacrene-D, β-caryophyllene, humulene, δ-cadinene and γ-cadinene. A sesquiterpene alcohol, elemol, is also present (3).

Among the sesquiterpenes, 15 furanosesquiterpenes are the typical substances of myrrh. Their skeletons correspond to furanoeudesmane, furanoelemane, furanogermacrene and furanoguaiane. Furanosesquiterpenes are easily identified by colour reaction with *p*-dimethylaminobenzaldehyde in methanol and concentrated HCl. A red-violet colour arises.

The main substance of the essential oil (19%) is furanoeudesma-1,3-diene (**1**) (6). It is a colourless, aromatic liquid, optically active (α_D -59°) and very instable, not described until now in the literature. In the mass spectrum the base peak at m/e 108 is characteristic of furanosesqui-

Margaris N, Koedam A, and Vokou D (eds.): Aromatic Plants: Basic and Applied Aspects
ISBN 90-247-2720-0.
Printed in the Netherlands.

terpenes with an unsubstituted carbon atom in α-position to the furan. In the ^{1}H-NMR-spectrum the α-furan proton has the resonance signal at δ=7.06; the protons of the β-methyl group appear at δ=1.95. In the IR-spectrum the furan causes vibrations at 1595, 1565, 1145 and 1090 cm^{-1}.

The tricyclic carbon skeleton and the number of olefinic carbon atoms were elucidated by ^{13}C-NMR-spectroscopy. The intensive "out of plane" bond at 725 cm^{-1} in the IR-spectrum is characteristic of an 1,2-*cis*-disubstituted olefin.

The exact position of the double bonds was determined by comparison with the NMR-spectrum of gazaniolid, isolated by Bohlmann and Zdero (1). We could not detect any Nuclear-Overhauser-Effect between the methyl groups at carbon atoms 14 and 15. Therefore, the C-15 methyl group must be fixed at C-10 and not at C-5.

Furanoeudesma-1,3-diene (1) was accompanied (6) by the earlier described lindestrene (2), which is also a colourless, instable aromatic liquid. The two compounds were separated over silicagel impregnated with silver nitrate. Until now lindestrene is found only in two plant species of the Lauraceae. Its structure is elucidated by Takeda et al.(7).

From the polar fractions of the essential oil we could isolate in small quantity another liquid furanoeudesmene with aromatic smell. According to the spectroscopic data it must be furanoeudesma-1,4-diene-6-one (3), not described until now (6). The ^{13}C-NMR spectra indicate the presence of one carbon atom with an oxo-group. The resonance in the unusual high field indicates an α,β,α',β'-olefinic conjugation of the oxo-group. Furthermore the significantly reduced carbonyl frequence in the IR-spectrum suggests also an α,β,α',β'-conjugation. The base peak of the Retro-Diels-Alder fragment at m/e 122 is characteristic of compounds with an oxo-group in α-position to the furan. Its position could be at C-6 or C-9. In agreement with the NMR-spectra of some known sesquiterpenes the oxo-function could only have the position at C-6. The two double bonds which are conjugated to the 6-oxo group must be tetrasubstituted. This is in agreement with the Δ^{7}-double bond of furan and with the Δ^{4}-double bond in the remaining skeleton.

From the group of furanoelemenes we have isolated isofuranogermacrene (4) and its 6-oxo-derivative curzerenone (5). Both furanosesquiterpenes are liquid with an aromatic smell. Takeda's group (5) and Hikino et al.(4) had isolated these two substances from one plant species of Lauraceae and one of Zingiberaceae, respectively, and have elucidated their structures.

The sesquiterpenes of the elemene group can be easily identified with ^{1}H-NMR: the free protons of the vinyl group (1-H, 2a-H, 2b-H) form an ABX system; its X-part (1-H) produces a characteristic double doublet resonance at δ=5.90 (J_{AX}=18 Hz; J_{BX}=10.5 Hz). The characteristic signals of the furan group are mentioned already.

From all furanosesquiterpenes, the furanoelemenes have the lowest concentration in the oil of myrrh.

Nine of the isolated furanosesquiterpenes belong to the type of furanogermacrene, viz. 1(10)E,4E-furanodiene (6) with its 2-methoxy and 2-acetoxyderivatives (6a, 6b), 1(10)E,4E- and 1(10)Z,4Z-furanodiene -6-one (7, 8), 4,5-dihydrofuranodiene-6-one (9) with its 2-methoxy- and 5-acetoxy-2-methoxyderivatives (9a, 9b) and 3-methoxy-10-methylene-furanogermacra-1-ene-6-one (10) (2). In the ^{13}C-NMR spectrum the ten membered ring is recognised by the ten distinct lines, which are in good resolution.

The known furanogermacrenes are mainly characterized by *trans*-configurated double bonds in 1(10)4-position. Only five furanogermacrenes of the essential oil of myrrh show this configuration; Two of them are 1(10)E,4E-furanodiene (6) and 1(10)E,4E-furanodiene-6-one (7), mentioned already in the literature (3). To confirm the structure of an isolated 1 1(10)E,4E-furanodiene we carried out a Cope-rearrangement (3) which is very characteristic for furanodiene. This reaction has a great stereoselectivity. From the stereochemistry of the reaction product, the starting compound can be identified. Pyrolysis of furanodiene at 160^{0}C (Fig. 1) leads to isofuranogermacrene (4). This demonstrates that the isolated furanodiene has the configuration 1(10)E,4E. Its 2-methoxy- and 2-acetoxy-derivatives are solid compounds not referred previously in the literature. We also isolated a new *cis*-isomer of furanodiene-6-one with 1(10)Z,4Z configuration. This configuration was verified by means of the Nuclear-Overhauser-Effect. 1(10)Z,4Z-furanodiene-6-one (8) is a liquid with a strong smell, while 1(10)E,4E-furanodiene-6-one is solid (6).

Three of the furanogermacrenes, isolated for the first time, are dihydro-compounds with a double bond in 1(10)-position. Until now, these dihydro-furanogermacrenes had not been found in nature (6). 4,5-Dihydrofurano-diene--6-one (9) is liquid while its 2-methoxy- and 5-acetoxy-2-methoxy-derivatives are crystalline compounds (3). Only the 5-acetoxy-2-methoxy-4,5-dihydro-furano-diene-6-one has a bitter taste but it is not so strong as that of myrrh (6).

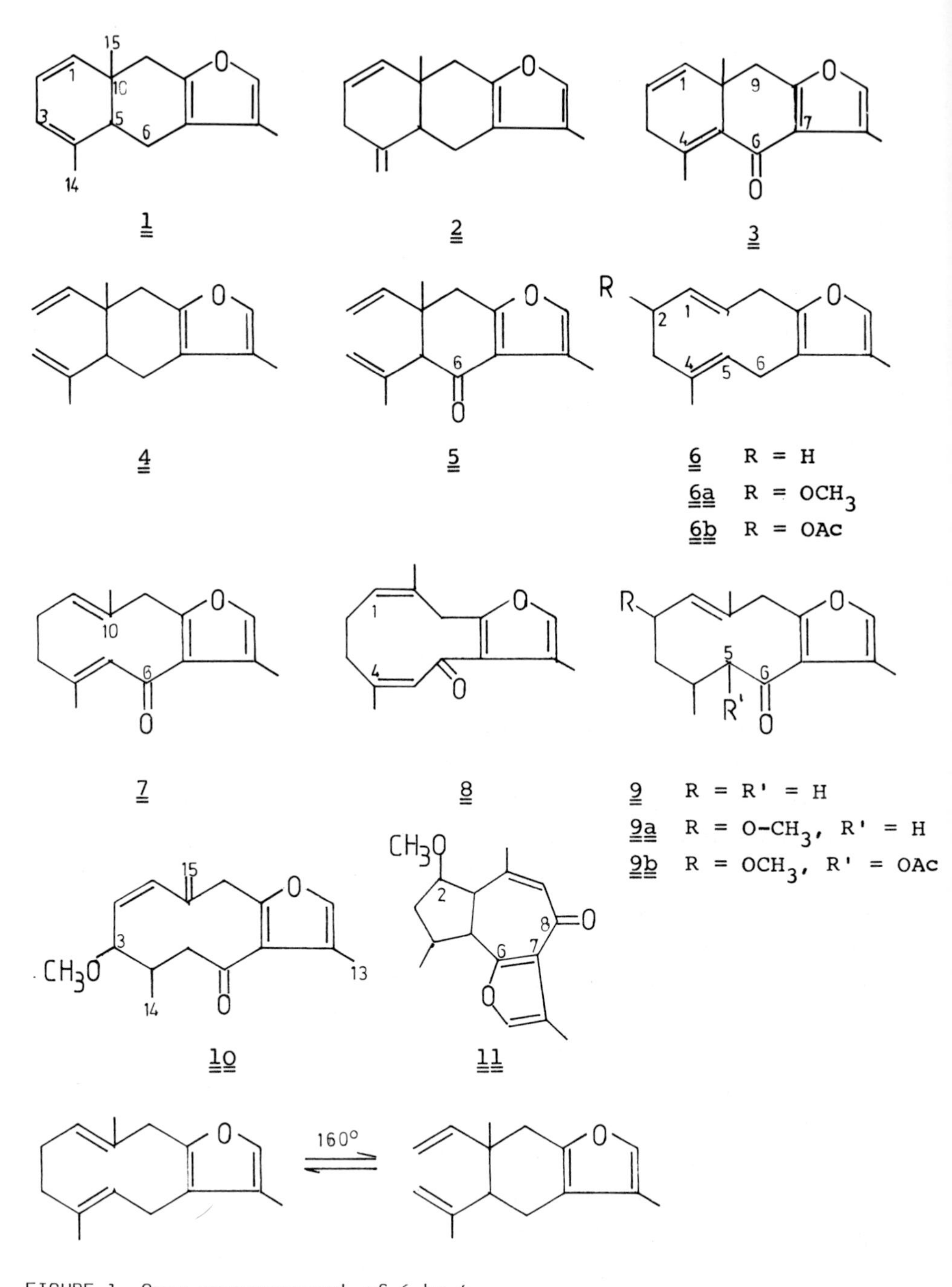

FIGURE 1. Cope-rearrangement of 6 to 4.

In order to ascertain whether the acetoxy group is fixed at C-5 in (**9b**) we carried out an NMR-shift experiment with the shift reagent $Eu(fod)_3$ (6). By this experiment we could indicate that the C-6 oxo group has the strongest complex acting qualities. For the geminal proton at the C-5 acetoxy group, vicinal to the C-6 oxo group, we have measured the greatest shift. Therefore, the position of the acetoxy group must be vicinal to the oxo group at C-6. By double resonance experiments we could make sure that shielded proton is the one at C-5 and not an olefinic proton.

The third new dihydrofuranogermacrene, 3-methoxy-10-methylene-furanogermacra-1-ene-6-one (**10**) (α_D +123°), is a liquid with a strong smell. It is a furanogermacrene with an exocyclic double bond not known until now in nature (3). In the ^{13}C-NMR-spectrum-off-resonance decoupled there are two doublets at δ=135.0 and δ=132.1, which belong to the carbon atoms of a 1,2 disubstituted double bond. A triplet at δ=115.6 and a singlet at δ= 142.5 indicate the exomethylene group between C-10/C-15. Therefore, (**10**) compared with furanodiene (**6**) has only two quartets for methyl groups (C-14, C-13) at δ=18.9 and 9.1. Following the UV-maximum at 238 nm the two double bonds which do not belong to the furan must be in conjugation. The determined value is characteristic of a heteroannular diene, arranged in s-*trans* conformation. By $LiAlH_4$ reduction we ascertained that the position of the exomethylene group is at C-10. By catalytic hydrogenation we received two compounds. One of them was the 3-methoxy-4,5-dihydrofurano-diene-6-one; its isomer, the 2-methoxy-4,5-dihydrofurano-diene-6-one (**9a**) is, as reported, a component of the essential oil of myrrh. The other was the 3-methoxy-tetrahydrofuranodiene-6-one.

2-Methoxyfurano-guaia-9-ene-8-one (**11**) is the second furanosesquiterpene of guaiane type, found in nature (6). Previously, only guidion from a plant of the Thymeleaceae was known. The new furanoguaiane is present in a quantity of 3% of the total oil. It is colourless with an aromatic smell, and instable. Its structure was elucidated by ^{13}C-NMR and ^{1}H (90 MHz) NMR-spectroscopy.

The isopropyl group at C-7 could build the furan at C-6 or C-8; both possibilities are known in the guaianolide chemistry. However in the guaianolide type the ring is not a furan but a γ-lactone. According to the spectra received the furan in the isolated furanoguaianolide (**11**) is only possible between C-7 and C-6. A 500 MHz ^{1}H-NMR-spectrum was a further proof for this assumption.

The presence of different types of furanosesquiterpenes in one plant though biogenetically possible is not frequently encountered. Precursor of all these compounds is supposed to be the germacra-1(10),4-diene-cation (Fig. 2). Transformation to the elemene type could take place by a Cope-rearrangement.

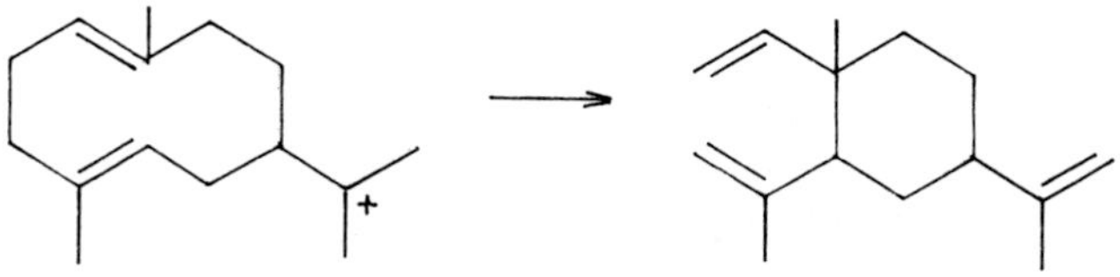

FIGURE 2. Rearrangement of germacra-1(10),4-diene-cation to elemene.

Based on these results we could easily separate various types of myrrh by TLC. (Stationary phase: SiO_2, mobile phase: dichloromethane running up to 10 cm. Compounds were detected by dimethyl-aminobenzaldehyde and hydrochloric acid).

REFERENCES

1. Bohlmann F, Zdero C. 1979. *Phytochemistry* 18:332.
2. Brieskorn CH, Noble P. 1980. *Tetrahedron Letters* 21:1511.
3. Brieskorn CH, Noble P. 1982. *Planta Med.* 44:87.
4. Hikino H, Agatsuma K, Takemoto K. 1968. *Tetrahedron Lett.* 2855.
5. Ishii H, Tozyo T, Nakamura M, Takeda K. 1968. *Tetrahedron* 24:625.
6. Noble P. 1980. Inhaltsstoffe des ätherischen Öls der Myrrhe. Neue Furanosesquiterpene vom Eudesman-, Germacran- und Guaiantyp. Ph.D.Thesis, Univ. of Würzburg.
7. Takeda K, Minato H, Ishikawa M, Miyawaki M. 1964. *Tetrahedron* 20:2655.

CHAPTER 5

Production and Application

THE INFLUENCE OF SOME DISTILLATION CONDITIONS ON ESSENTIAL OIL COMPOSITION

A. KOEDAM

1. INTRODUCTION

The development of gas chromatography has accelerated research activity in the field of essential oils. Thus, after the introduction of this technique there has been almost an exponential increase in the rate of publication of papers concerned with the chemical composition of essential oils. The vast majority of these studies was conducted on oils obtained from the plant material by means of distillation. Thus far little attention has been attributed to the influence of this method of isolation on the composition of the oils. For instance, isomerization and saponification of the more labile compounds may take place depending upon distillation conditions. In consequence, the results of GC analysis will not present the actual composition of the oil in the living plant, but only the chemical composition of the isolated product. In studies on the influence of seasonal variation or geographical location on plant constituents any transformation due to the distillation method may have serious implications upon the ultimate conclusions. Nevertheless, information on this part of isolation procedures is scattered and generally buried under other results. Although some data may be derived from a review on production and quality control of essential oils by Garnero (2), it appears that no systematic examination on this important feature has been undertaken.

This paper was prepared with the aim to give a report on two aspects of distillation that may exert a severe influence on the composition of essential oils. First, the length of the distillation period will be discussed, and, secondly, attention will be focussed on the acidity of the distillation water.

2. RESULTS AND DISCUSSION

2.1. Influence of the distillation period

The influence of the distillation period is demonstrated with the volatile seed oil from dill (*Anethum graveolens* L.). Two components predominate

Margaris N, Koedam A, and Vokou D (eds.): Aromatic Plants: Basic and Applied Aspects
 ISBN 90-247-2720-0.
Printed in the Netherlands.

in this oil: the monoterpene hydrocarbon limonene and its oxygenated analogue carvone. Together they represent ca 95% of the oil.

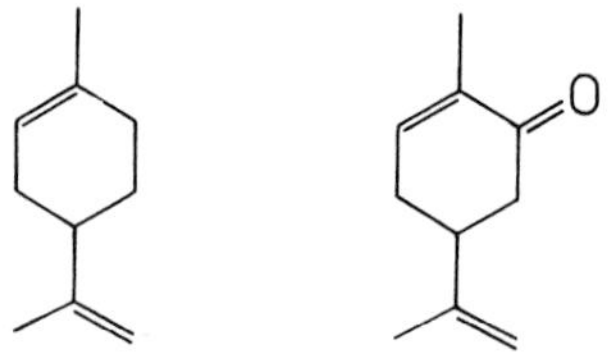

limonene
(bp 176 °C)

carvone
(bp 230 °C)

Oil batches obtained after various distillation times showed considerable differences in composition. Fig. 1 gives gas chromatograms of dill oil obtained after distillation for 1 h and after 6 h. As will be seen after 1 h the distillate consists mainly of carvone (ca 90%), while only a small amount of limonene is present. Comparing this with the composition after 6 h a striking difference is noticed. The proportion of limonene is greatly increased and approaches 35%. It was found that the amount of limonene in the oil gradually increases as the distillation proceeds, whereas the amount of carvone steadily diminishes. The changing ratio of limonene and carvone with the time of distillation is visualized in Fig. 2.

From this graph it will be seen that the percentage of carvone in the oil is slowly declining with an attendant increase of that of limonene. It is also evident that it takes at least 12 h before the ratio becomes fairly constant.

To get more information on the backgrounds of this phenomenon, subsequent experiments were performed in which the distillate was collected in separate fractions during sequential intervals of 30 min. GC examination of these fractions provided a more detailed picture of the altering composition with the time of distillation. The results are given in Fig. 3. During the first 75 min the fractions consisted for the greatest part of carvone; thereafter limonene dominated. The fractions collected after 150 min contained only small amounts of carvone. After 180 min this compound could no longer be detected: further fractions were only made up of limonene. From these data it must be concluded that carvone distills before limonene. However, if the boiling points of both compounds are considered - 176 °C for limonene and 230 °C for carvone - this conclusion does not seem to be very obvious.

A search of the literature revealed that this improbable behaviour was

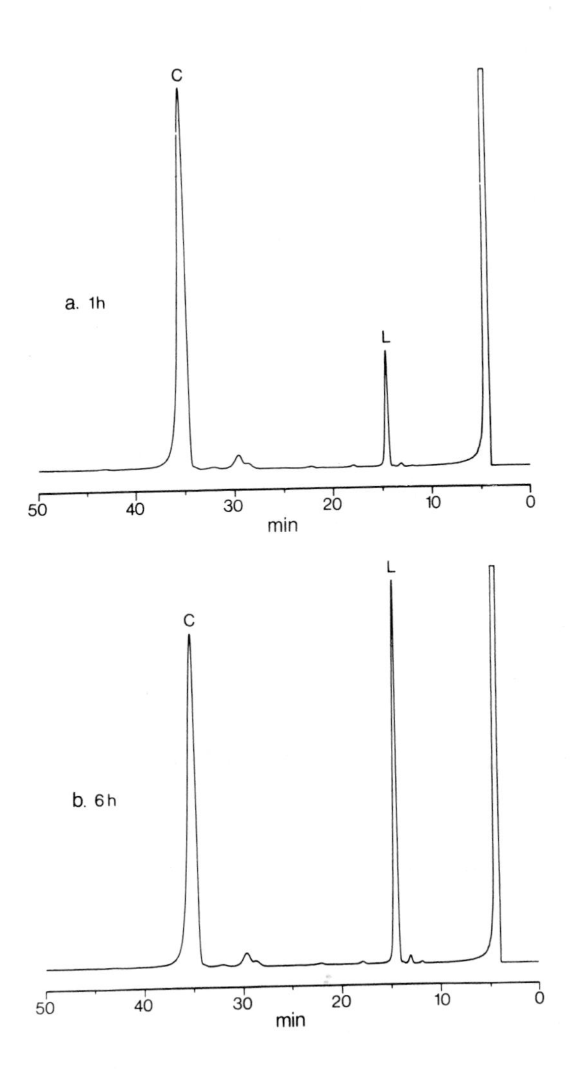

FIGURE 1. Gas chromatograms of dill oil obtained after different distillation times. L = limonene; C = carvone.

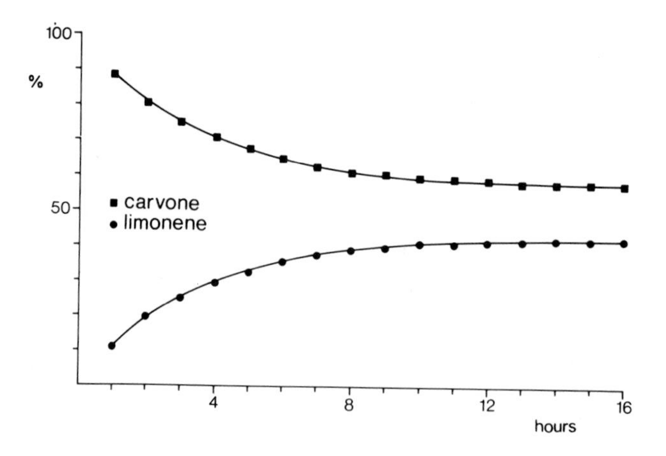

FIGURE 2. Relative amounts of limonene and carvone in dill oil as a function of the time of distillation.

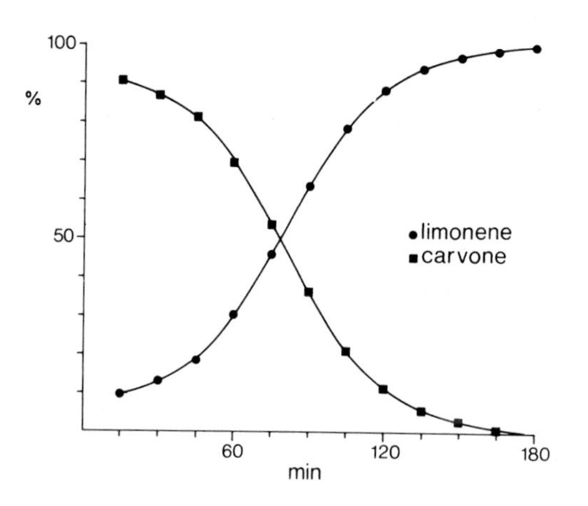

FIGURE 3. Relative amounts of limonene and carvone in sequential fractions of dill oil.

already described in the middle of the last century by Zeller (10) who noted that during distillation of caraway oil the less volatile oxygenated fraction distilled before the hydrocarbons, but the author could not find a satisfactory explanation. Several decades later a fundamental work on the physical backgrounds of essential oil distillation was compiled by Von Rechenberg (7). Unfortunately, his "Theorie der Gewinnung und Trennung der ätherischen Öle durch Destillation", published in 1910, did not attain the attention it deserved, which was undoubtedly prevented by the fact that it was never translated into English. However, in his book Von Rechenberg gives an accurate elucidation of the phenomenon which reads as follows: during distillation the boiling water penetrates the plant tissues and dissolves a part of the essential oil present in the glands. This aqueous solution diffuses through the cell membranes. Once arrived at the surface the oil is immediately vaporized. This process continues until all the enclosed oil is removed from the cells. Since the oxygenated constituents are much better soluble in the boiling water than the hydrocarbons, diffusion of the first is highly favoured. Thus, during distillation of essential oils from intact plant material the rate of diffusion of the different compounds through the cell membranes is of greater importance than the boiling points of the components and, therefore, the compounds vaporize according to their degree of solubility in the distillation water rather than following the order of their boiling points. This means that the higher boiling oxygenated compounds, being polar - hence more soluble - will distill before the hydrocarbons. This type of diffusion with water as the carrier was called hydrodiffusion by Von Rechenberg; it is the driving force behind the feature described above.

2.2. Acidity of the distillation water

The oil obtained from leaves of the savin juniper (*Juniperus sabina* L.) will serve as example to deal with the effect of the acidity of the distillation water on the composition of the distillate.

One of the main constituents of this oil is the monoterpene hydrocarbon sabinene (ca 30%). When oil sampled after different distillation times was analyzed, it appeared that the amount of sabinene was steadily declining, whereas at the same time the percentage of an oxygenated compound, terpinen-4-ol, slowly increased.

The transformation of sabinene into terpinen-4-ol under the influence of diluted mineral acids was already described in the first decade of this century by Wallach (8). More recent investigators (1, 4, 5) confirmed the rearrangement of sabinene, and reported, in addition to terpinen-4-ol, also α-terpinene, γ-terpinene and terpinolene as reaction products (Fig. 4). As it was observed in our experiments that the amounts of these substances increased simultaneously with the decrease of sabinene, it was assumed that, due to the acidity of the distillation water, a certain extent of conversion occurred. Thus it seemed worthwhile to investigate the influence of the pH of the distillation water during isolation of the oil. For this purpose

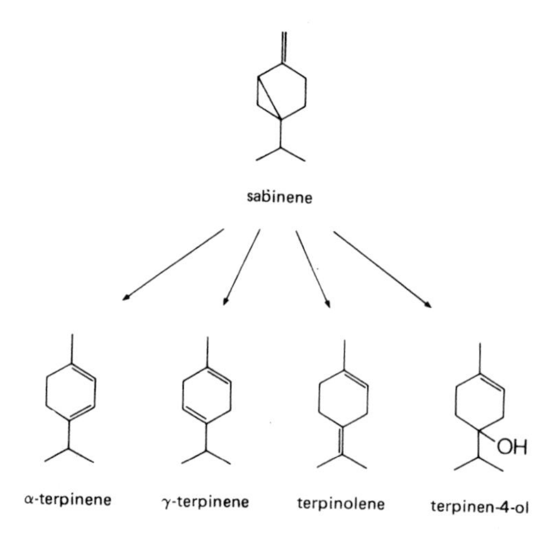

FIGURE 4. Rearrangement of sabinene under acidic conditions.

distillations were performed with plant material that was immersed in buffer solutions covering the pH range 2.2-8. The results are reported in Fig. 5. As is evident the proportion of sabinene in the oil differs widely over the investigated pH range. At low pH values the amount of sabinene in the distillate was about 7.5%. When the solution is rendered alkaline, however, this percentage shows a sharp increase to nearly 40% at pH 8. The inverse is observed for the transformation products α- and γ-terpinene, terpinolene and terpinen-4-ol. The levels of these compounds are gradually reduced when the pH value of the water is raised.

Structurally closely related to sabinene are both *cis*- and *trans*-sabinene hydrate. They are also not very acid-proof: rearrangement leading to terpinen-4-ol may proceed easily, as was originally observed by Wallach (9). In recent years this conversion was extended by Granger and coworkers (3) and by Taskinen (6) reporting also α- and γ-terpinene as products. From Fig. 5 it will be seen that both sabinene hydrates slowly decreased as the pH of the water was lowered; they were even no longer detected at pH values below 4.

Based on these findings it must be concluded that the acidity of the water accounts for the production of α-terpinene, γ-terpinene, terpinolene and terpinen-4-ol at the expense of both sabinene and the sabinene hydrates. To our opinion conversion of these compounds during distillation is quite common. In fact, its occurrence may also be derived from data reported by several authors, although it has not been recognized as such.

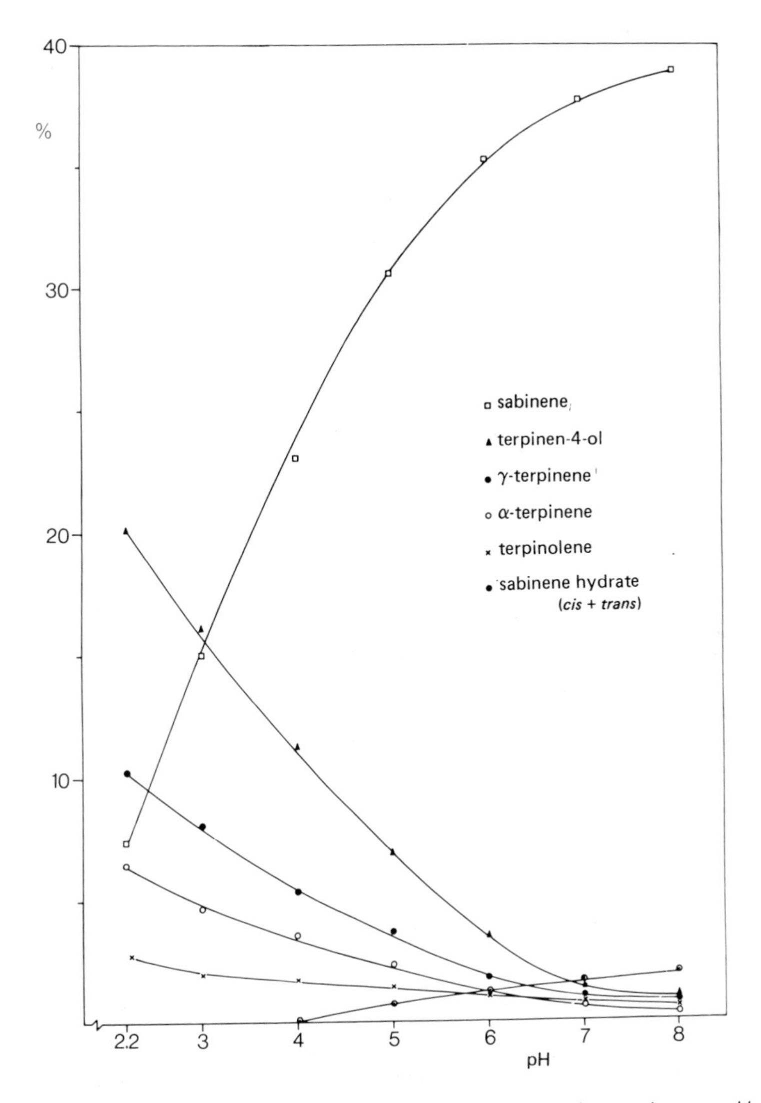

FIGURE 5. Influence of the acidity of the distillation water on the percentages of some components in savin juniper oil.

3. CONCLUSION

During distillation essential oils are liberated from the structures where they are stored in the plant by diffusion. This diffusion can be rather time-consuming, as is the case with umbelliferous seeds, since the oil must pass the hard seed cover. Furthermore if labile compounds are present, such as sabinene and its hydrates, a high degree of rearrangement may be observed, depending upon the acidity of the distillation water.

It has to be noted that solvent extraction of plant material will avoid both diffusion and acidity and, therefore, in some cases has to be preferred over the distillation method.

In studies on the influence of external factors on the composition of essential oils not only environmental circumstances must be exactly recorded, but also the subsequent analysis must be sufficiently accurate, without giving rise to changes in the oil constituents. This aspect has been treated by many research workers, and analysis of essential oils by means of GC has substantially improved.

However, another factor which must be taken into consideration is the isolation of the oils from the plant material. This procedure should be harmless, as it is the intention to study the genuine composition of the oils. This factor has often been neglected. Thus, in many papers a biosynthetic correlation in the terpinene-terpinolene-terpinen-4-ol group, in close association with sabinene, is indicated. The majority of these studies was performed with oils obtained by distillation. However, in view of our results, the effect of different isolation conditions must not be overlooked. The examples given above clearly illustrate that distillation can have a remarkable influence on the composition of essential oils.

REFERENCES

1. Cooper MA, Holden CM, Loftus P, Whittaker D. 1973. *J. Chem. Soc. Perkin Trans. II* 665.
2. Garnero J. 1976. *Rivista Italiana EPPOS* 58:105.
3. Granger R, Passet J, Lamy J. 1975. *Rivista Italiana EPPOS* 57:446.
4. Norin T, Smedman L-Å. 1971. *Acta Chem. Scand.* 25:2010.
5. Tolstikov GA, Lishtvanova LN, Goryaev MI. 1963. *Zh. Obshch. Khim.* 33:683.
6. Taskinen J. 1976. *Int. Flavours Food Addit.* 7:235.
7. Von Rechenberg C. 1910. Theorie der Gewinnung und Trennung der ätherischen Öle durch Destillation, pp. 418-441, Miltitz bei Leipzig, Selbstverlag von Schimmel & Co.
8. Wallach O. 1907. *Ber. Dtsch. Chem. Ges.* 40:585.
9. Wallach O. 1908. *Justus Liebigs Ann. Chem.* 360:82.
10. Zeller GH. 1851. *Jahrb. Prakt. Pharm. Verw. Fächer* 22:292.

VARIATIONS IN YIELD PARAMETERS IN A WILD POPULATION OF *ORIGANUM VULGARE* L.*

E. PUTIEVSKY, U. RAVID

1. INTRODUCTION

Oregano (*Origanum vulgare* L.), family Labiatae is a perennial species which grows wild in hilly regions, mainly in the vicinity of the Mediterranean Sea and in Central America. Under semi-cultivated conditions it is grown in Turkey, Morocco, Western Europe, Mexico, and Argentina. In those areas oregano is harvested mainly from wild populations, once or twice a year. The dried leaves are used extensively in pizzas, tomato dishes and others, as well as in the food industry. The essential oil is distilled from the leaves and utilized widely as a flavouring agent in sauces, soups and liqueurs (11).

Oregano is known in two main types: Mexican and European. The Mexican oregano is of the genus *Lippia* and is used mainly in Central America and southeastern USA. Oregano from Greece is known as the best of the European types.

In wild populations and even in cultivated fields, there is large variation in plant morphology, leaves, and essential oil characteristics (5). To the best of our knowledge, not much has been done to develop a uniform cultivated variety after screening variations in wild populations.

In our experiments with cultivated oregano in Israel (10), considerable variation was found in various characteristics. In order to select the most suitable lines for cultivation in Israel tests were carried out with representative oregano types on the following: morphological characteristics, essential oil content and concentration of the major components, and leaf yields. This paper reports the results of two years of observations.

2. MATERIAL AND METHODS

Seeds of oregano (*Origanum vulgare* L.) received from fields near Lamia (Greece) were sown in a greenhouse in Israel on 15 Nov. 1973. Seedlings

*Contribution from the Agricultural Research Organization, The Volcani Center, Bet Dagan, Israel. No. 196-E, 1981 series.

Margaris N, Koedam A, and Vokou D (eds.): Aromatic Plants: Basic and Applied Aspects

ISBN 90-247-2720-0.
Printed in the Netherlands.

with ten leaves were transplanted into experimental plots at the Newe Ya'ar Experiment Station on 20 Feb. 1974 in rows spaced 25 X 50 cm (10). In 1977, after three years of growth and observation, ten plants representing the range of variation in the field, were chosen for further experiments. These plants (lines), forming a separate plot, were propagated vegetatively by stem cutting. In January 1977 the cuttings were planted in rows 40 X 15 cm, in plots 2.5 m X 10 m plots of each line. Ammonium nitrate fertilizer, 300 kg ha^{-1}, was applied before planting and after each harvest. From the end of March until November, the plots were irrigated weekly with 250 m^3 ha-l of water.

Morphological data were recorded during 1977 and 1978 before harvest. For each character, 15-20 random samples from each line were tested. Each line was harvested four times per year in April, June, August and October.

All harvesting was performed with a motorized cutter. Four samples (1X1 m) were taken from each line, weighed and dried. The leaves and stems were separated manually. The content of volatile oil in the dried leaves from the four samples of every line was determined with a Clevenger apparatus.

Samples of the essential oils were analyzed by gas-liquid chromatography. Identification of the essential oil main components and determination of their percent content, were carried out on a Packard 7400 GLC provided with a spectrophysics "System 1" integrator. Analysis was carried out using an 8 X 1/8" glass column packed with 10% Carbowax 20M on CW AW 80-100 mesh. The operating conditions were as follows: initial column temperature, $80^{o}C$, programmed at a rate of $5^{o}C$ min^{-1} to $190^{o}C$; inlet temperature, $250^{o}C$; detector oven temperature, $230^{o}C$; carrier gas: nitrogen; flow rate, 15 ml min^{-1}; hydrogen flow rate, 30 ml min^{-1}, air flow rate 300 ml min^{-1}. Similar conditions were previously found to be optimal for oregano oil analysis (2).

3. RESULTS

The optimal time for propagation of oregano was found to be early autumn (9). Planting was done in winter, because at that time it is very easy to detect differences between plants. The late planting produced a relatively low yield in the first year of growth (Figs. 1, 2).

The morphological differences among the ten lines are represented in Table 1. Plant death observed after establishment and two harvests was

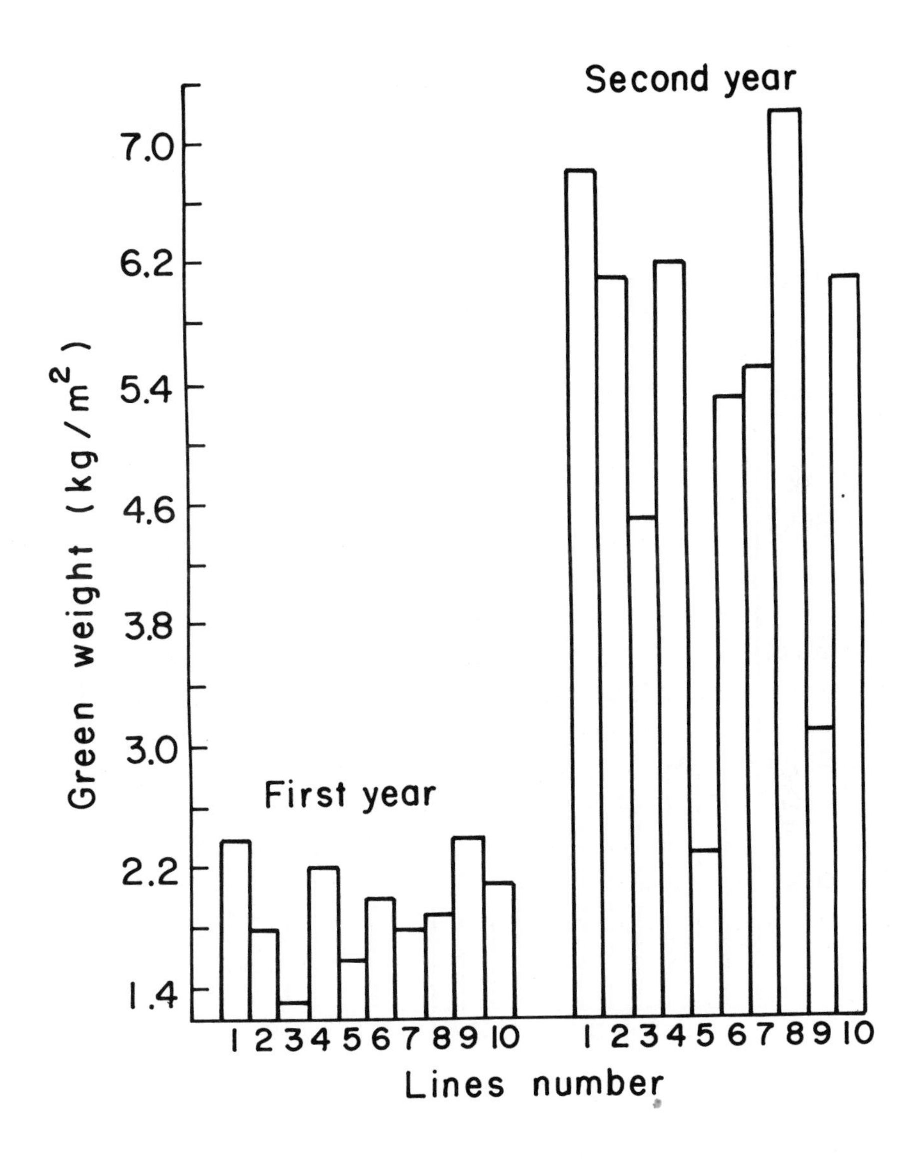

FIGURE 1. Yield of the ten lines (fresh leaves) during two successive years.

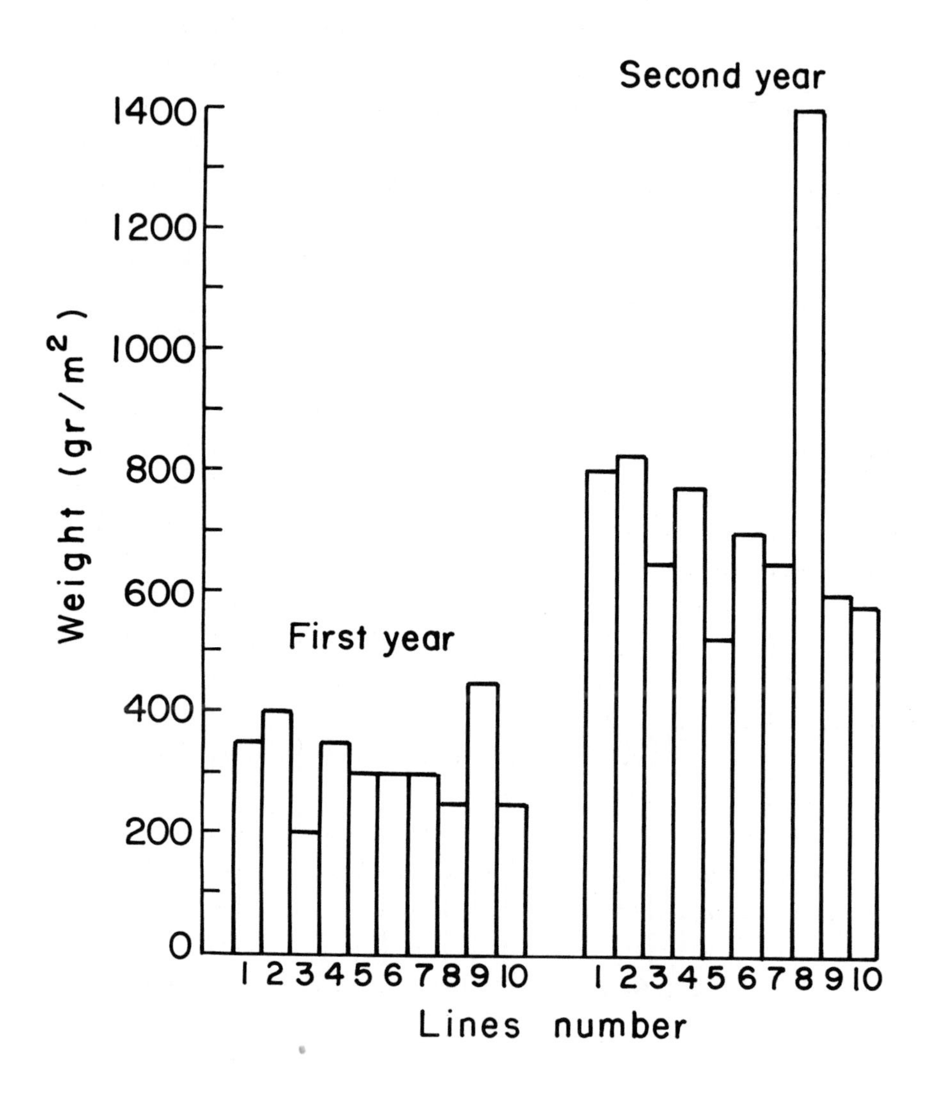

FIGURE 2. Yield of the ten lines (dry leaves) during two successive years.

high (33%) in lines 3, 5 and 6, whereas in lines 1 and 9 no plants died. Flowering was observed during July in lines 1, 4, 7, 8 and 10, but none of the other five lines flowered in that month. The number of branches per plant was extremely high in line 2 (80) and only one-quarter (22) of that maximum in lines 8 and 10. The internode variation was also quite high, reaching the maximum length in lines 2, 4, 7 and 8, and the minimum in line 9. Leaves showed extremely large variations, especially as to the surface area: the biggest leaves (lines 3 and 5) were almost seven times larger than the smallest ones (lines 4, 7, and 8). There was no positive correlation between the leaf surface area and the number of leaves. Moreover, the number of leaves differed on the two measurement dates. In spite of this, line 4, with the smallest leaves, had many leaves in April and in October, whereas line 6 - also with small leaves - had relatively few leaves in both cases.

The length and width of the leaves were measured twice, in April and October (Table 1). Concerning with the leaf size, great differences were observed in the same line between April and October but also among different lines in each of these months. A high correlation was found between leaf surface area, number of leaves per plant, and length and width of the leaves, in October. For example, line 4, with many leaves (126) and a small leaf surface area (0.4 cm^2), had a small leaf size (10X5 mm); line 5, with quite many leaves (105) and a big leaf surface area (2.88 cm^2), had large leaves (20X16 mm).

On the basis of morphological characters, the ten lines could be divided into two main groups: one including lines 4, 7, 8 and 10, and the other including lines 3, 5, 6 and 9; lines 1 and 2 are intermediate. The first group did not flower in July (including line 1), there were only few branches (including line 1), the internodes were long (including line 2), the number of leaves was low (except for line 4), leaf surface area was small (except for line 10), and individual leaf size as well as the l/w ratio were large (Table 1).

The fresh plant weight per year increased from the first to the second year (Fig. 1). Lines 1 and 10 gave the highest yield and lines 5 and 9 the lowest; all the other lines were intermediate. The dry leaf yield was high in line 8 and low in line 5 (Fig. 2). Line 2 was consistently better than all the other lines during the two years; line 9 gave high both fresh and dry yields only in the first year of growth.

Table 1. Some morphological characteristics of ten Oregano lines in the first year of growth.

Line No.	Plant death (%) in July	Flowering in July	Number of branches per plant in July	Internode length (cm) in October	Leaves: Number per plant in April	Leaves: Number on main branch (10cm) in October	Leaves: Size (of biggest leaf), mm, in April: Surface area (cm^2)	in April: Length	in April: Width	in April: Ratio (l/w)	in October: Length	in October: Width	in October: Ratio (l/w)
1	0_{a}	yes$_{a}$	35_{ab}	1.8_{a}	90_{b}	100_{b}	1.40_{b}	18_{a}	16	1.13_{a}	15_{ab}	11_{ab}	1.36_{a}
2	28_{b}	no$_{b}$	80_{c}	2.7_{b}	85_{b}	85_{ab}	1.77_{b}	26_{b}	16	1.63_{b}	13_{ab}	7_{a}	1.86_{b}
3	32_{b}	no$_{b}$	51_{b}	1.7_{a}	65_{a}	96_{a}	2.32_{c}	26_{b}	20	1.30_{a}	23_{b}	16_{b}	1.44_{a}
4	24_{b}	yes$_{a}$	56_{b}	2.7_{b}	150_{c}	126_{b}	0.40_{a}	21_{b}	15	1.40_{ab}	10_{a}	5_{a}	2.00_{b}
5	35_{b}	no$_{b}$	60_{b}	1.5_{a}	80_{b}	105_{b}	2.88_{c}	25_{b}	22	1.14_{a}	20_{b}	16_{b}	1.25_{a}
6	33_{b}	no$_{b}$	40_{ab}	1.6_{a}	110_{c}	141_{b}	1.14_{b}	22_{b}	17	1.29_{a}	15_{ab}	10_{ab}	1.50_{a}
7	10_{a}	yes$_{a}$	38_{ab}	3.0_{b}	30_{a}	56_{a}	0.65_{a}	18_{a}	15	1.20_{a}	9_{a}	5_{a}	1.80_{b}
8	8_{a}	yes$_{a}$	22_{a}	2.6_{b}	75_{b}	63_{a}	0.89_{a}	25_{b}	14	1.79_{b}	15_{ab}	8_{a}	1.88_{b}
9	0_{a}	no$_{b}$	26_{a}	1.3_{a}	150_{c}	82_{ab}	1.88_{b}	27_{b}	20	1.35_{a}	20_{b}	15_{b}	1.33_{a}
10	6_{a}	yes$_{a}$	22_{a}	2.0_{ab}	50_{a}	75_{ab}	1.72_{b}	25_{b}	16	1.56_{ab}	12_{a}	7_{a}	1.71_{b}

Figures followed by different letters differ significantly at $P<0.05$

There were great differences in the essential oil content among the ten lines. In Fig. 3 there is the essential oil yield (expressed as percentage of the dry leaf weight) in all four harvests in the second year of growth, and in Fig. 4 the total yield of essential oil in both years. The highest level of essential oil was obtained in summer (second harvest - July), and the lowest in autumn (fourth harvest - October). Only three lines did not follow this pattern; in lines 7 and 9, the content rose from harvest to harvest, while in line 10 the maximum content was recorded in the third harvest. The essential oil content varied from 2.8% to 6.4% in dry leaves in the different harvests and lines.

The yield in essential oil (Fig. 4) was quite low (10 ml m^{-2}) in the first year. There were large differences among the lines; i.e. line 9 yielded 12 ml m^{-2} and line 4 only 5 ml m^{-2} ($P<0.05$). In the second year the maximum yields of essential oil were obtained from line 2 (55 ml m^{-2}) and line 8 (50 ml m^{-2}) and the lowest yield from line 4 (35 ml m^{-2}), as in the first year.

Gas chromatographic analysis of the essential oil showed that thymol was the major component in all ten lines (Table 2). Even so, lines 4, 7, 8 and 10 had a low thymol content (15-20%), lines 2, 3, 6 and 9 a high thymol content (60-70%), and lines 1 and 5 intermediate (44-47%). A relatively high carvacrol content (6-7%) was obtained in lines 1 and 5, and a low level in lines 6 and 9 (1.2-1.6%). *p*-Cymene level varied in the ten lines from 16% in line 6 up to 35.7% in line 4; γ-terpinene was high (9%) in lines 4, 7, 8 and 10.

4. DISCUSSION

An attempt was made to obtain some information about variation, as a basis for selection, in a wild population of oregano. All the parameters examined varied in one degree or another, but from the agricultural and economic points of view the important variations are those in yield, leaf characters and essential oil content. Both dry leaf weight and essential oil content were in some lines almost the double of others; however, the increased values of these characters did not coincide in one and the same line. To obtain the two characteristics in one line, an effort must be made in artificial crosses, or to find lines with male sterility (5, 8) in order to use this phenomenon to achieve natural hybridization. However, it should be remembered that only one population was examined, and it is possi-

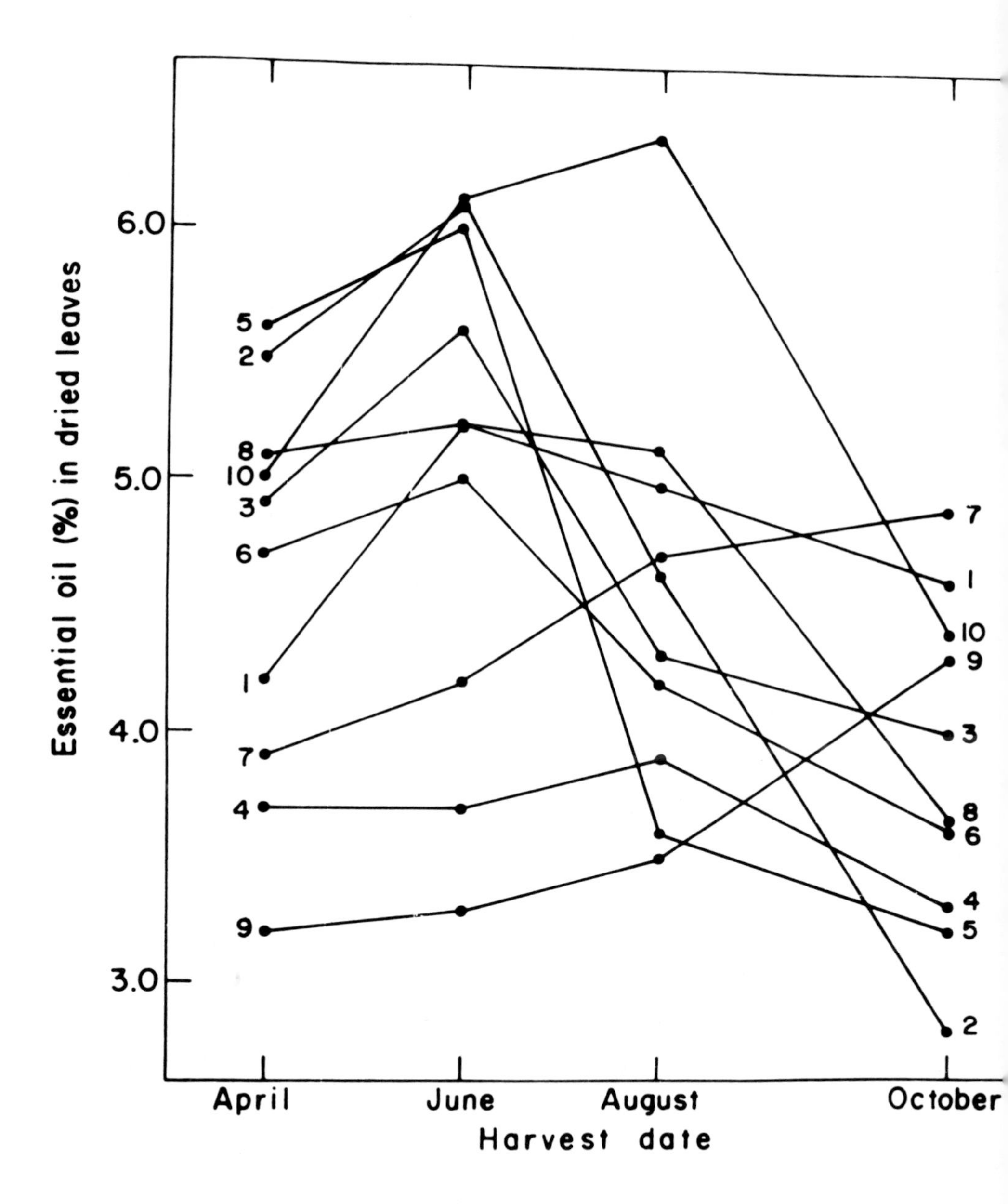

FIGURE 3. Essential oil content in ten oregano lines during four harvest times.

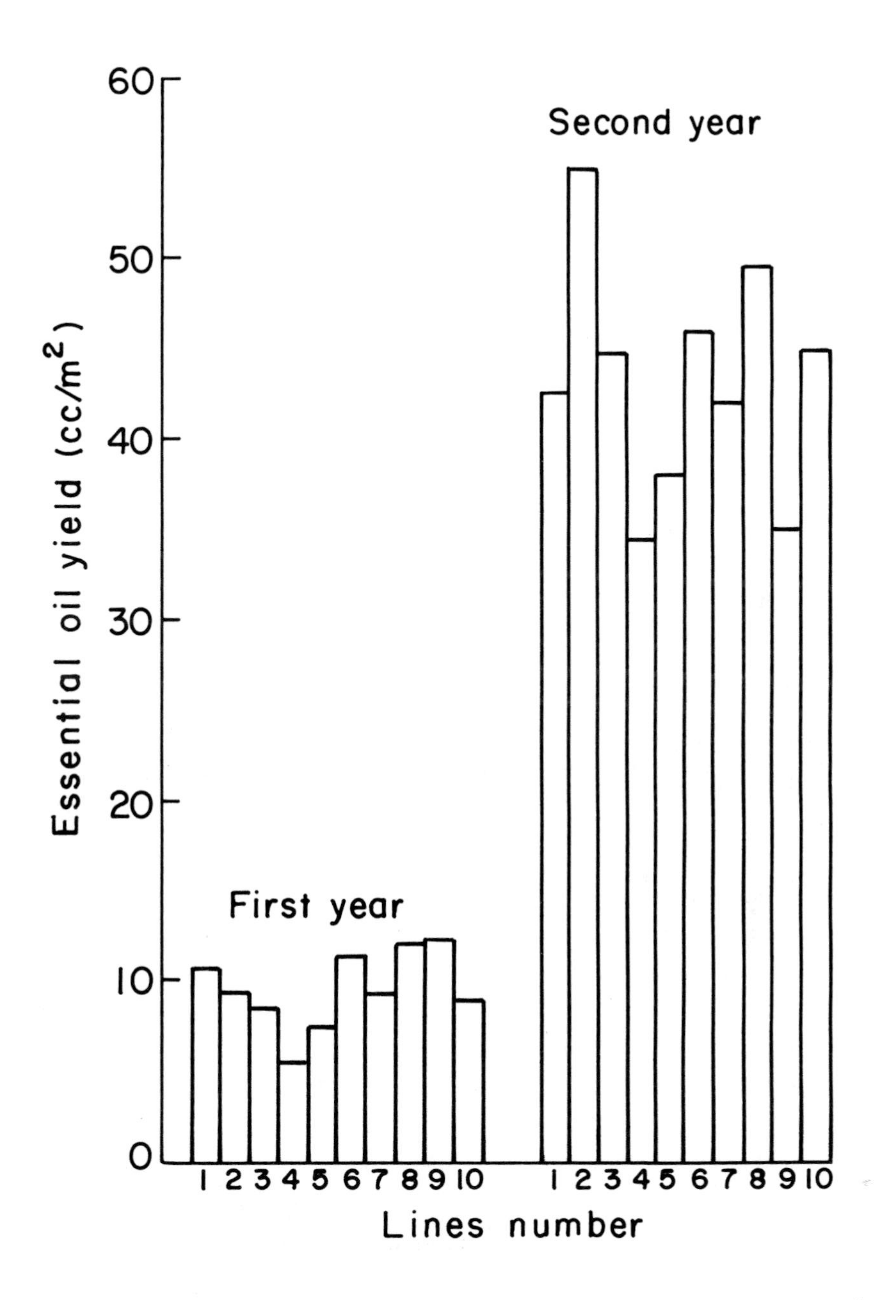

FIGURE 4. Essential oil yield of ten oregano lines during two successive years.

Table 2. Main components of the essential oil of ten lines of Oregano in spring (April) harvest.

Line	Percentages of the major components of Oregano essential oil			
No.	Thymol	Carvacrol	p-Cymene	γ-Terpinene
1	43.8_{ab}	7.8_{b}	27.4_{b}	4.6_{a}
2	67.1_{b}	3.6_{ab}	18.5_{a}	4.6_{a}
3	68.2_{b}	3.3_{ab}	23.0_{b}	5.1_{a}
4	15.2_{a}	4.7_{ab}	35.7_{b}	9.3_{b}
5	46.7_{ab}	6.7_{b}	30.0_{b}	4.8_{a}
6	70.8_{b}	0.6_{a}	16.0_{a}	4.2_{a}
7	18.6_{a}	2.4_{a}	21.3_{ab}	10.5_{b}
8	19.5_{a}	4.4_{ab}	21.7_{ab}	9.1_{b}
9	61.7_{b}	1.2_{a}	26.5_{b}	5.6_{a}
10	19.8_{a}	4.8_{ab}	21.5_{ab}	9.5_{b}

Figures followed by different letters differ significantly at $P<0.05$.

ble that the best combination of characteristics might be found in individuals from other populations. For different purposes (leaves or essential oil), it would be useful to have different lines.

For the sake of brevity, not all data are presented here. Nevertheless, the data on essential oil content (Fig. 3) show that at some harvests the level was quite high, while at others it was low. Hence, it is possible to harvest plants of the same line on one date for dried leaves, and on another date as a source of essential oil.

Variations among lines were very high, and the ten lines could be divided into three groups. In the first group are lines 4, 7, 8 and 10, which have many similar morphological characteristics, as well as similar essential oil content. Concerning their essential oil, they are of the "marjoram" type, with relatively low content of thymol and carvacrol. This group has long and compact inflorescences, few leaves in the flowering branches, and an almost complete absence of petal leaves (Figs. 5, 6). Line 1 has some characteristics common to the first group and to the

FIGURES 5 and 6. Branches, each from different line at full blossom. The line numbers are from the left to the right.

second group, which includes lines 2, 3, 5 and 9. This second group has similar morphological characteristics; similarity is extended to the major essential oil components. The variation in the ten lines could be considered as a continuous one that presents all the range in this population. This variation gives indirect evidence that this species has a genetic system of open pollination, or a system of incompatibility (3, 4).

Oregano may be a very good subject for selection, using the possibility of propagation from seed in order to obtain the maximum variation, and to preserve this variation or successful lines by propagation from cuttings 6, 7). It would be very easy to produce a uniform field by stem-cutting propagation. The question arises whether uniform fields have an advantage, not only from the agricultural point of view, but also from the economic point of view. For intensive agriculture it is of great importance to have uniform plants that respond identically to herbicides, fertilization, irrigation, harvest dates and methods, etc. However, the market's demand is not uniform and it is therefore important to have mixed populations, or a range of varieties with different flavours (1). It may be possible to solve these problems by selection of different varieties for different purposes and needs.

REFERENCES

1. Anonymous. 1980. Markets for Spices in North America, Western Europe and Japan. International Trade Center, GATT, Geneva.
2. Ashkenazy A. 1977. Content and composition of essential oils in *Mentha piperita* and *Origanum vulgare* and their influence on feasibility of growing these plants in Israel (in Hebrew). M.Sc. Thesis, Tel-Aviv University, Ramat Aviv, Israel.
3. Clifford J. 1976. Introduction to Natural Selection. Baltimore, University Park Press.
4. Falconer DS. 1961. Introduction to Quantitative Genetics. New York, Ronald Press Co.
5. Jain SK. 1968. *Nature* 217:764.
6. Kuris A, Altman A, Putievsky E. 1981. *Scient. Hort.* 13:53.
7. Kuris A, Altman A, Putievsky E. 1981. *Scient. Hort.* 14:151.
8. Lewis D, Crowe LK. 1956. *Evolution* 10:115.
9. Putievsky E. 1978. The optimal date for planting oreganum (in Hebrew). *Hassadeh* 58:1269.
10. Putievsky E, Basker D. 1977. *J. Hort. Sci.* 52:181.
11. Rosengarten F. 1969. The Book of Spices. Philadelphia, Livingstone Publ. Co.

THE PRODUCTION OF AROMATIC PLANTS IN THE PANCALIERI AREA

F. CHIALVA

1. INTRODUCTION

In view of the general revaluation which aromatic plants are obtaining throughout the world, partly as a result of the greater ecological awareness which has been awakened in recent years, it is worth mentioning the production of Pancalieri, a small village in Piedmont which, although not important from a quantitative point of view, undoubtedly represents a very special and isolated case in the production of aromatic plants and non-citrus essential oils in Italy.

Aromatic plant production, a tradition which is now centuries old, is managed by a few companies which work almost full-time on it and a large number of small family businesses which devote themselves part time and only to a limited extent to this activity. The land is flattish and divided up into small areas (Fig. 1). The part which is not reserved for aromatic plant cultivation is occupied by traditional crops (maize, wheat, grassland) and by nursery gardens of fruit and timber trees.

2. PANCALIERI PROFILE

Pancalieri (44^0 50′N latitude, 7^0 34'E longitude) is situated at an altitude of 241 m above sea level covering a surface of 1597 ha of which 1196 ha is agricultural land; its population is approximately 2000 inhabitants. Data on the average temperature and rainfall are in Fig. 2 and on sunshine in Fig. 3. The distribution of the population in different economic activities is shown in Fig. 4 and agricultural land use pattern in Fig. 5.

3. PEDOLOGY

The recent alluvial soil, free from limestone (pH 6-6.7) is particularly suited to the type of crops which do not tolerate an asphyxiated environment. The good drainage given by the sandy nature of the layer for cultivation and the sand and gravel combination of the substratum, together

Margaris N, Koedam A, and Vokou D (eds.): Aromatic Plants: Basic and Applied Aspects

ISBN 90-247-2720-0. Printed in the Netherlands.

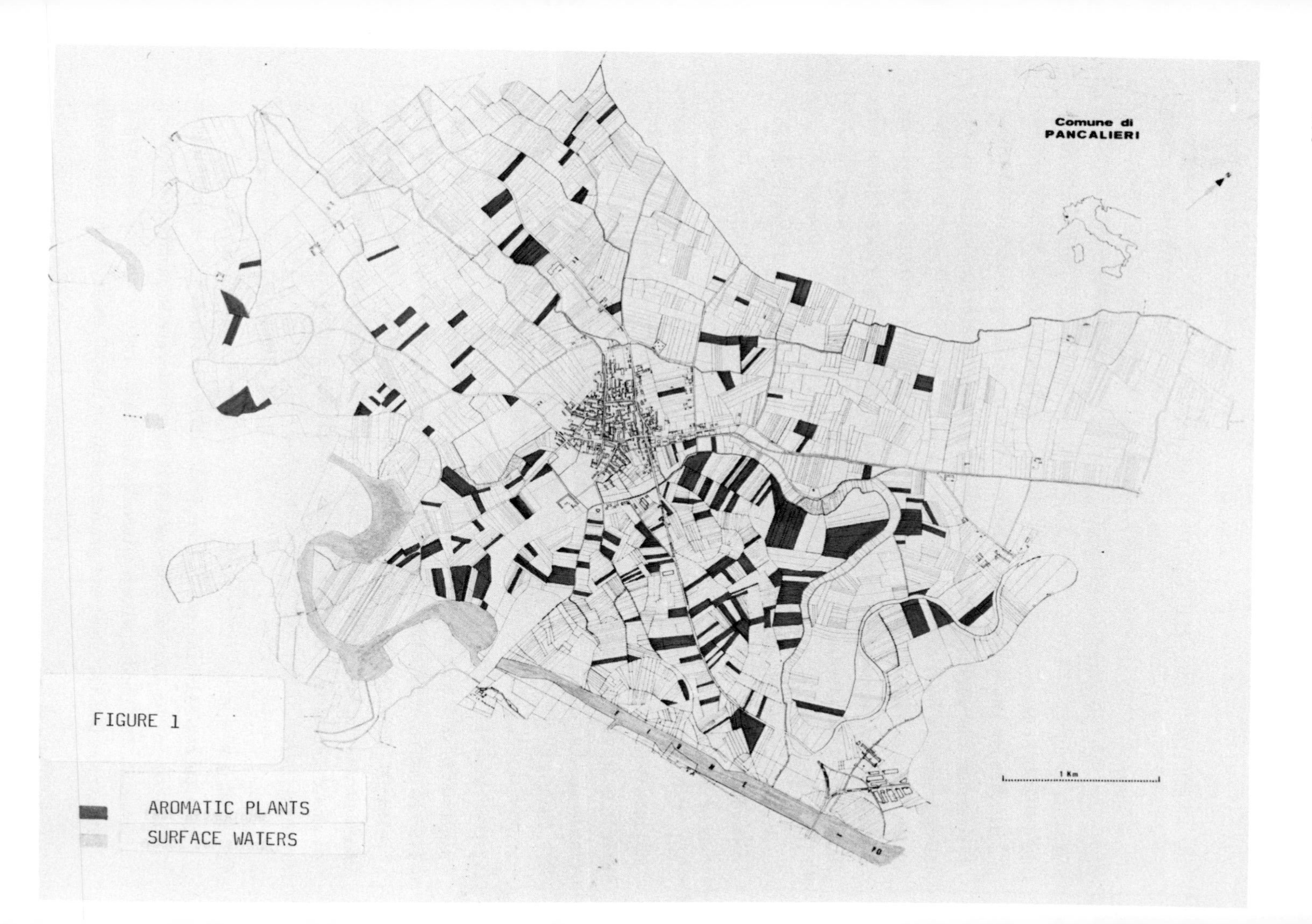
Comune di
PANCALIERI
1 Km
PO
AROMATIC PLANTS
SURFACE WATERS

FIGURE 1

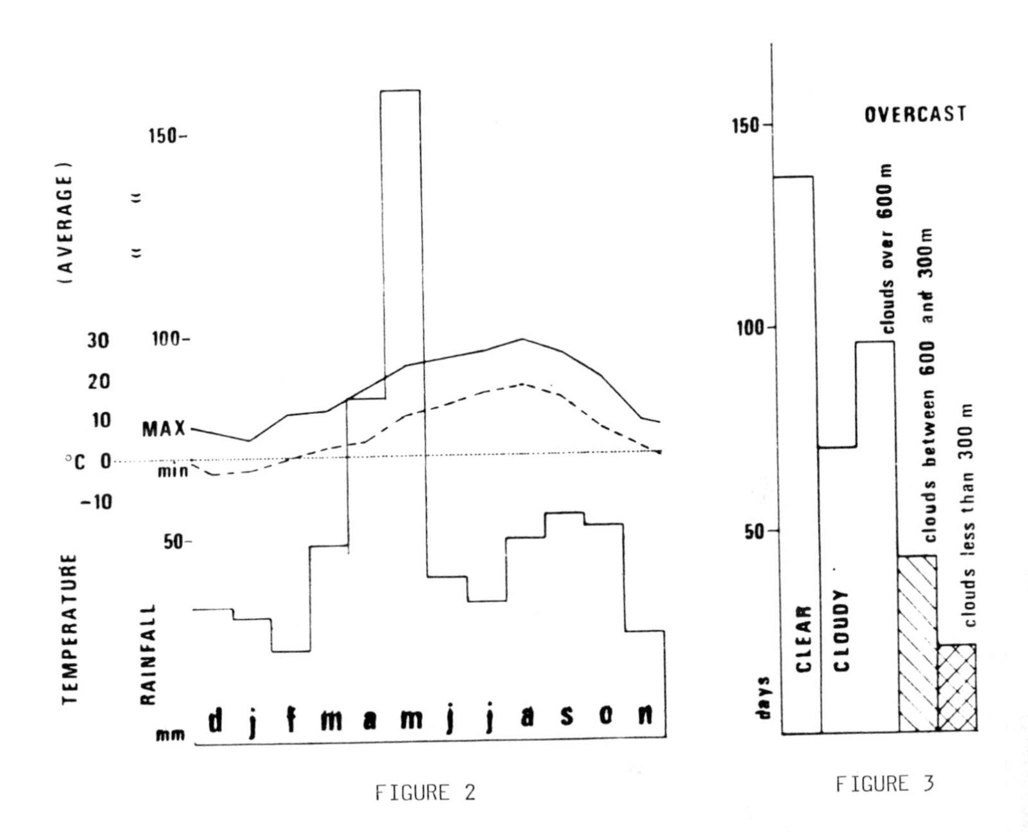

FIGURE 2

FIGURE 3

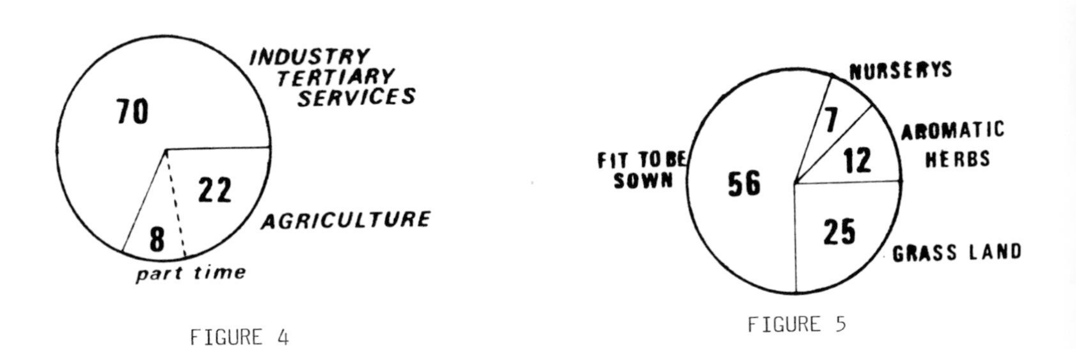

FIGURE 4

FIGURE 5

with the abundance of water-bearing strata close to the surface and the proximity of the river Po, are the principal factors which contribute to the natural fertility which characterises this zone. The constant bringing of organic material and large scale fertilisation determine moreover optimum growth and development conditions for these particularly demanding crops.

4. CULTIVATION

Almost all the area cultivated with aromatic plants is covered by Roman wormwood, mint and tarragon; the first is marketed in the form of dried herb and the other two as essential oils. A small portion of this area, only a few hectares, is cultivated with St. John's wort, marjoram, melissa, hyssop, savory and Roman camomile. The latter, until a few years ago, occupied a considerable part (up to 20 hectares) but difficulty in placing the product caused the progressive abandonment almost to the point of disappearance.

Roman wormwood (*Artemisia pontica* L.). About 50 ha are grown. The production is around 200 tons of dried herb destined for the drink industry and is marketed in Italy.

Tarragon (*Artemisia dracunculus* L.). Production occupies about 60 ha with a yield of about 3000 kg of essential oil. The oil, used in the alimentary and perfume industries is almost all exported. There is also a small demand for dried leaves destined for health food shops because of the beneficial properties for stomach and digestion which the plant posesses.

Italo-Mitcham peppermint (*Mentha piperita* Huds. var. *officinalis* Sole f. *rubescens* Camus). Pancalieri is rightly considered the historic centre of mint cultivation in Italy, but the area occupied has been diminishing year by year; it is estimated that in 1981 about 25-30 ha, are cultivated yielding 1500-2000 kg of essential oil. In spite of the considerable competition of mints of other origins, the essential oil of mint produced in Pancalieri is considered the best in the world for alimentary use in the confectionary, drink and pharmaceutical industries, by virtue of the delicacy of its flavour and the pleasantness of its taste. These particular characteristics are derived essentially from the reciprocal relationship of certain constituents of the oil; these characteristics are obviously linked with the plant's biochemical nature which is conditioned by particular pedoclimatic factors.

5. HOW THEY ARE GROWN

The type of soil preparation is common to all these plants. It consists of deep ploughing in the spring, harrowing to ensure complete breaking up of the sods, and abundant fertilisation with phosphorus and potassium salts (1-1.2 metric tons ha^{-1} of complex dressing type 8:24:24, for example). Before transplantation a bland chemical weedkiller can be used as a precautionary measure, especially for mint, but not for absinthe. For transplantation, carried out with a special semi-automatic transplanter devised on the spot, young plants are used, which are derived from the springtime regrowth of the previous year's stump. These plants are obtained vegetatively and are opportunely prepared right from the autumn. The transplantation ratio is of about 1:10 for tarragon, 1:18 for mint and 1:30 for absinthe. The period of transplantation goes from the second half of April for tarragon to the first half of May for Roman wormwood, with a common planting density of about 150,000 young plants ha^{-1}. While covering over nitrogenous fertiliser with urea or ammonium nitrate base is applied (0.3-0.5 metric tons ha^{-1}). After this, manual and mechanised weeding is carried out until the harvest. Possible irrigation, compatible with mint, is very harmful to tarragon and absinthe.

6. HARVEST

Tarragon is distilled in the middle of July, immediately before flowering and mint is distilled in the first days of August, when the plant is in full bloom. Distillation is conducted in a steam current and some measures have been taken in order to facilitate and accelerate the loading and unloading of the still. The waste material is redistributed on the land and covered with soil; it does not have great fertilizing properties but it possesses excellent restoring qualities.

Roman wormwood is harvested in the first days of August; after cutting, the green herb is transported to the factory and made into small bundles which are left to dry out on wires stretched under the roofs of the houses. When drying is complete, the bundles, joined together in groups, are piled up and kept until the moment of sale.

7. MECHANISATION

The need to make up for the shortage of labour and to reduce the hard physical work of the employees, together with the precise awareness of the necessary operations of cultivation led the operators to look for alternative solutions and better technological efficiency. The operators themselves have devised transplanting machines and equipment for mechanical weeding which are particularly suitable for use with these crops, some of which, like tarragon, are very delicate and their young plants are particularly fragile. The loading and unloading operations of distillation have been improved too, and speeded up in the course of these last years, by introducing some innovations (like the cold pre-loading), which facilitate the employées'work considerably. The making of bundles for the natural drying of absinthe in the shade is still done entirely by hand, although experiments are already being made, with good results, with artificial drying carried out with air at a relatively low temperature (37-40°C).

SOME POTENTIALLY IMPORTANT INDIGENOUS AROMATIC PLANTS FOR THE EASTERN SEA-BOARD AREAS OF SOUTHERN AFRICA*

S.R.K. PIPREK†, E.H. GRAVEN, P. WHITFIELD

1. INTRODUCTION

The Ciskei Essential Oils Project (CENTOIL project) was established in 1974 in an effort to discover, investigate and develop labour intensive crops which could contribute meaningfully to the development of the agricultural resources and the uplift of the tribal communities of rural Ciskei.

The Ciskei lies between 32°33′ and 32°17′S Latitude and 27°09′ and 28°10′E Longitude, and forms part of the Eastern Seaboard of Southern Africa.

The area earmarked for the production of essential oils consists of a dissected coastal plane which rises to an altitude of 1500 m at a distance of 60 km from the sea. The climate is in the main warm temperate but varies considerably over relatively short distances. The rainfall varies from less than 500 mm in the valleys to more than 1000 mm in the mountains and occurs mainly in the summer and autumn. The region is rich in indigenous plant species and has a plentiful supply of labour which will respond willingly to economic incentives.

Peppermint (*Mentha piperita*) appears to have some economic potential for the area and it is envisaged that one or more of the species dealt with in this paper may suitably complement it. One of the species (*Pteronia*) is a pernicious weed in the area and holds the added attraction of possibly converting a current liability into a revenue-earning asset.

2. LANYANA (*Artemisia afra*)

Lanyana, also known as Wildeals, or Wormwood, is an indigenous aromatic shrub that is fairly abundant in the Amatola and Drakensberg mountains of Southern Africa. It is an erect growing, woody perennial with pinnatisect leaves. Under favourable growing conditions in its natural habitat it can

---

* Contribution from Department of Agronomy, University of Fort Hare, Alice, South Africa.

† Deceased July 1979

Margaris N, Koedam A, and Vokou D (eds.): Aromatic Plants: Basic and Applied Aspects

ISBN 90-247-2720-0. Printed in the Netherlands.

attain a height of 150 cm. Lanyana is one of the oldest known Southern African medicinal shrubs and was formerly used by the Hottentots and subsequently by the settlers as a tonic and as an anthelmintic, usually in the form of tinctures (4).

Following a detailed anatomical study, Botha and Herman (1) reported that the oil bearing ducts in *Artemisia afra* are spatially associated with the primary vascular bundles in the young stems and with the primary veins of the petioles, petiolules and leaflets.

The essential oil from Lanyana has an attractive blue-green colour and has been equated with good quality Armoise oil for which an established world market already exists. Using conventional steam distillation procedures the first and predominant oil fraction has a light green colour which becomes progressively bluer as the distillation proceeds. The last small fraction has a dark blue colour. The oil content of plants in the vegetative stage varies from 0.3% to 0.9%, depending upon the growth stage and growing conditions experienced. The seed is particularly rich in oil and yields of 1.4% have been obtained from the fine residue of seeds and leaves collected from the barn floor where Lanyana is stored prior to distillation.

The major constituents of Lanyana oil are listed in Table 1.

2.1. Production Aspects

Collection of wild material in the mountains by local tribesfolk is subject to variations in supply and quality due to shortage of labour at certain times of the month, the vagaries of the climate, the stage at which the plants are harvested, forest fires, etc. There is also reason to doubt whether sufficient material is available from this source to support a significant industry.

The production of Lanyana as a conventional arable crop appears to offer a solution to most of the problems listed and a research programme was initiated in 1975 to investigate this possibility. In brief, it was found that Lanyana seed has a definite light requirement for germination* and that the plant, once established, grows vigorously and can compete very effectively with local weeds - to the extent that the possibility of the

---

*Investigation by P.P.J. Herman, Department of Botany, University of Fort Hare, Alice, South Africa.

Table 1. Major Constituents of Lanyana (*Artemisia afra*) Oil*.

Constituents identified	Percentage
Terpinolene	0.51
α-Pinene	0.30
γ-Terpinene	0.75
Camphene	1.15
p-Cymene	0.66
1.8-Cineole	13.00
α-Thujone	52.50
β-Thujone	13.10
Camphor	6.55
Artemisia Ketone	1.96
Sesquiterpene-1	1.62
Sesquiterpene-2	1.49
Sesquiterpene-3	1.38
TOTAL	90.48

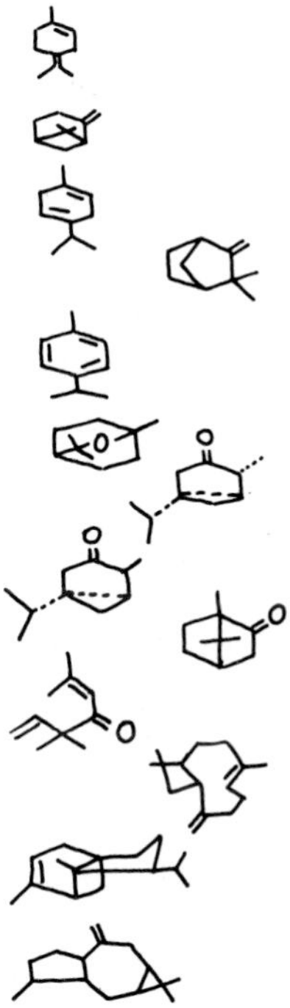

plant releasing an allelopathic substance which naturally inhibits the growth of other plants is currently being investigated. It would appear that provided a sufficiently attractive economic return can be assured, little difficulty will be experienced in persuading the local tribesfolk to replace their rather meagre crops currently growing on the steep mountain slopes, with Lanyana which in fact occurs as a weed in the area. Lanyana as a crop, also has the advantages that it is a hardy plant not readily eaten by livestock and is less prone to pests and diseases than the conventional food crops currently produced. It can also play an important part in combating soil erosion.

*Analysis provided by Dr. E.E.Bartels, Department of Chemistry, University of Stellenbosch, Republic of South Africa.

It has been observed that the wild Lanyana purchased from the tribesfolk appears to give higher yields per ton of green material than the material produced under arable conditions. The differences are probably related to the degree of moisture content and also to the fact that drought coupled with certain "unfavourable" growing conditions may enhance the oil content of the "wild" material.

When grown as a domesticated crop, Lanyana is inclined to produce tall, woody stems, and shed a large proportion of its leaves before flowering in the autumn/winter. Distillation experiments have shown that the woody stems contain virtually no oil. Field experiments have indicated that if Lanyana is cut and distilled in the vegetative stage in mid-summer, it recovers rapidly and will produce a second growth that will attain full flower in the late autumn. In a cutting experiment during the 1975/76 season it was found that cutting Lanyana twice during the growing season resulted in a yield of 150 kg of oil per hectare as opposed to a yield of 50-70 kg of oil from plants which are allowed to proceed to maturity and are cut in autumn at which state they are very woody with few leaves. Provided care is taken to ensure that loss of leaf, flowers, and the brittle plant parts is carefully controlled, little oil loss occurs when the plants are dried naturally. Lanyana may be harvested, sun dried and stored for a considerable time before distillation without significant oil loss.

From the data presented in Fig. 1 it is apparent that sun drying of the material on plastic sheets for 240 hours after harvesting in winter has actually resulted in a slight increase in the oil yield, possibly because of a more effective movement of the steam through the plant mass in the still. In this experiment sun drying for 240 hours resulted in a 42% loss in mass and a concomitant reduction in volume. Drying after harvest and prior to distillation appears to have considerable merit from the point of view of harvesting at the correct stage for maximum yield and will also permit processing of a greater quantity of material per still which in turn will result in a higher yield of oil per still charge. This can materially influence the economy of the project by virtue of the savings in both fuel and labour.

It is also evident from Fig. 1 that little advantage is to be gained by increasing the distillation time to more than 90 minutes.

Depending upon the market price of the oil the Centoil Project is capable of producing substantial quantities of Lanyana oil and this crop holds

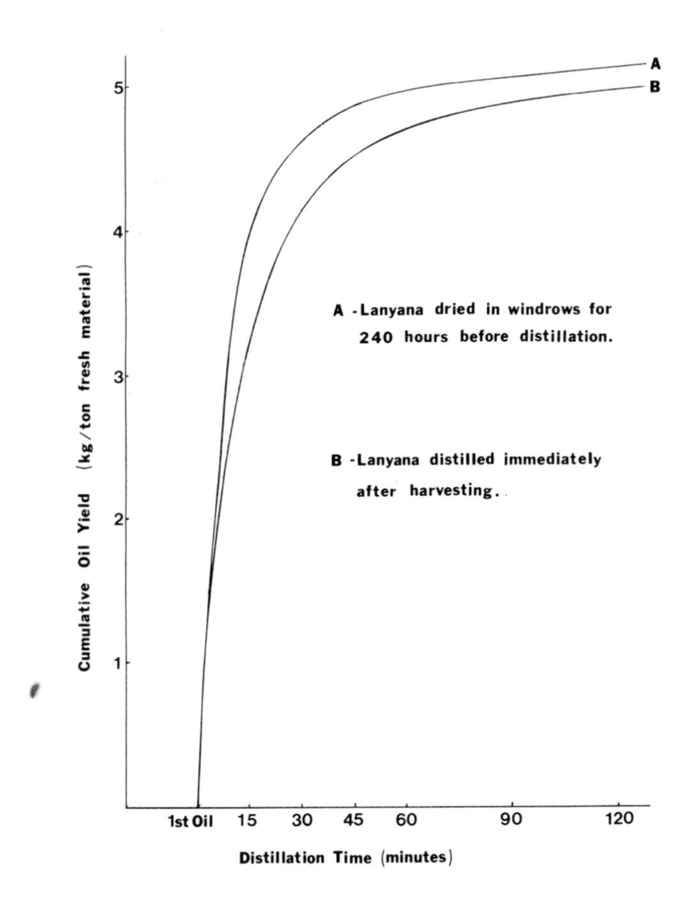

FIGURE 1. Effect of drying Lanyana prior to distillation on the cumulative oil yield over a two hour distillation period.

the prospect of materially contributing to the uplift of certain economically depressed communities.

3. PTERONIA (*Pteronia incana*)

Pteronia, also known as Blue Dog, is a hardy and aggressive, heath-like perennial shrub with small grey leaves and multi-branched stems. Under favourable conditions the plant reaches a height of about 100 cm. Both the leaves and the profusion of yellow flowers produced in spring and early summer, are strongly scented. Pteronia occurs in abundance in the semi-arid regions of the Ciskei where it is aggressively invading large areas of natural pasturage (veld). It has been estimated (5) that 60,000 hectares of land have been encroached by this weed in the greater Ciskei - this is a matter of concern to the local agricultural authorities.

The oil has a pale yellow colour and a fairly strong fragrance which has been adjudged by one expert as somewhat remotely resembling the essence of Juniper berries.

The major constituents of the oil are:

Pinene 14.2%; Pinene and Sabinene 29.8%;
Myrcene 17.7%; Cineole and Limonene 14%
p-Cymene 2.3% and Terpinolene 9.4%*

Where the harvested material has been permitted to dry out as a result of prolonged storage which also facilitates separation of the brittle leaves from the stems and twigs, it has been found that oil yields are reduced as compared to material distilled a few days after harvesting. It would also appear that the distillation period must be lengthened considerably in order to promote extraction of the oil from dry leaves. In one experiment better results were obtained where dry leaves were boiled in water during distillation as opposed to the standard procedure of steam alone passing through the material in the still. Distillation of the woody stems and twigs resulted in a low yield of dark yellow oil and appears to have little merit. Although the highest oil yields appear to be associated with an increase in the percentage of leaves and flowers, there appears to be no economic advantage in physically separating the two before distillation.

3.1. Production Aspects

The abundant supply of Pteronia available makes it unlikely that this plant will ever be cultivated as a conventional arable crop. It has been observed that where the top leaves, twigs and flowers of the plants are manually harvested with sickles, the plants recover rapidly and depending upon the rainfall experienced, can be harvested a number of times during the year. In the interests of improving the natural pasturage, the local agricultural authorities would welcome a system of destructive harvesting of the Pteronia plants.

From a study conducted in 1977, in which Pteronia was harvested at two-weekly intervals, it would appear that a more or less sustained supply of material can be harvested throughout the year with peak production occur-

*Analysis by Messrs Cavallier Frères, Grasse.

ring during the flowering season which extends from July to December.

Successful exploitation of the vast resources of Pteronia in the Ciskei is wholly dependent upon the development of a sustained and reliable market for the oil.

4. ERIOCEPHALÉE (*Eriocephalus punctulatus*)

Eriocephalée is a bushy, woody shrub up to 80 cm high, which is indigenous to certain mountainous areas of Southern Africa. The fleshy gland-dotted leaves have a particularly attractive fragrance and the plant was apparently used as a substitute for Buchu (*Agathosma* spp.) by native folk (4).

The essential oil of Eriocephalée has a deep blue colour, hence the name "essence plante bleue". The principal constituents of the colourless volatile fraction (boiling point 250°C) which constitutes 55% of the essence, have been reported by Roard et al.(3) (Table 2).

Table 2. Principal compounds in the volatile fraction of *Eriocephalus punctulatus* oil (3).

Compounds	Proportion of fraction %
2-Methylbutan-1-ol	0.50
3-Methylbutan-1-ol	0.10
2-Methylpropyl isobutyrate	14.00
α-Pinene	6.00
β-Pinene	2.00
2-Methylpropyl isovalerate	1.00
2-Methylbutyl isobutyrate	22.60
3-Methylbutyl isobutyrate	3.60
Limonene	1.00
p-Cymene	8.00
2-Methylbutyl butyrate	1.00
2-Methylbutyl isovalerate	4.50
2-Methylbutyl valerate	1.00
Terpinen-4-ol	2.00
α-Terpineol	0.50
	67.80

During the early 1970's limited quantities of Eriocephalée oil were produced and marketed by a South African based essential oil company. According to reports the oil was quite readily accepted by the perfumery trade and considerable optimism was expressed as to its commercial potential. With the commencement of commercial production it was found that the wild plants available could not supply sufficient material to meet the demand for the oil. Apparently, this was due to the initial overestimation of the amount of wild material available, coupled with the slow recovery growth of the plants after complete defoliation. Efforts to cultivate the plant as a conventional arable crop were not successful due to the difficulty of collecting viable seeds from the wild plants. These problems eventually led to the abandonment of commercial exploitation of this crop.

During 1980 the Centoil Project devoted considerable research attention to Eriocephalée. Using a Musashige and Skoog (2) medium with sodium diphosphate and various levels of auxin and cytokinin, it was found that shoot tip and nodal explants of Eriocephalée can generate new shoots and roots *in vitro*.*

Following these findings a practically feasible method of vegetative propagation has been developed. To date, a few hundred genetically identical plants have been produced for planting in the Amatola mountains. Although Eriocephalée does not occur naturally in the Ciskei mountains, it can be confidently expected that the plants will grow well in the same ecological areas where Lanyana occurs naturally.

Experiments relating to cutting schedules which will result in sustained yields of quality material will receive a high priority in the research program.

5. CONCLUSION

The successful establishment of reliable and substantial markets will be the prime factor determining the production of oil from each of the species discussed. Physical resources and technical expertise are available to produce substantial quantities of Lanyana and Pteronia provided the economic incentives are created. In the case of Eriocephalée which requires vegetative propagation in the glasshouse and which has a relatively slow initial growth rate, the establishment of substantial oil production

---

*Investigation by D. Hobkirk, Department of Agromony, University of Fort Hare, Alice, South Africa.

capacity will require somewhat more time.

REFERENCES

1. Botha CEJ, Herman PPJ. 1980. *Jl. S. Afr. Bot.* 46:197.
2. Musashinge T, Skoog F. 1962. *Physiol. Plant* 15:473.
3. Roard M, Derbesy M, Peter H, Remy M. 1977. *Int. Flavours Food Addit.* 8:29.
4. Smith CA, Phillips EP, Van Hoeffen E. 1966. Common names of South African plants. Pretoria, Govt. Printer.
5. Trollope WSW. 1975. In: The Agricultural potential of the Ciskei. Faculty of Agric. Mem. Fort Hare University, South Africa.

LABIATAE AS MEDICINAL PLANTS IN ISRAEL*

Z. YANIV, A. DAFNI, D. PALEVITCH

1. INTRODUCTION

The Labiatae family is known for the wealth of species with medicinal properties which have been used since early times (11). Many of these species are very common in the Mediterranean region.

There are 33 genera of the Labiatae growing wild in Israel, including 180 species (29). Many of them are rich in essential oils (12, 28). About ten species are noted in the literature as being used in folk medicine in Israel (4, 6). These species are common mainly in the mountainous areas of the Mediterranean parts of Israel.

No recent survey has been conducted until now on the ethnobotany of medicinal plants in Israel. This article presents some of the results of a survey on this subject carried out by the authors and still running with the purpose of finding potential medicinal plants which will be introduced as agricultural crops.

2. MATERIAL AND METHODS

The survey was conducted during 1980/81. Sixty selected informants were interviewed. Information regarding folk medicinal practice was collected for about 100 plants. The identity of the plants was checked by live specimens, photographs and slides. A medicinal property was accepted as valid if mentioned by at least three different informants. Most of the interviewed people are active as herbal healers, and their average age is 60 years. Healers who were popular and known in their area were chosen for the survey. In most cases only one healer from each village was interviewed.

3. RESULTS AND DISCUSSION

The results are presented in Table 1, which includes the species chosen and their most important medicinal properties in folk medicine in Israel.

The most important use, common to all of the surveyed species, is in the treatment of stomach pains and indigestion. *Salvia fruticosa* and

*Contribution from the Agricultural Research Organization, The Volcani Center, Bet Dagan, Israel. No. E-248, 1981 series.

Margaris N, Koedam A, and Vokou D (eds.): Aromatic Plants: Basic and Applied Aspects

ISBN 90-247-2720-0. Printed in the Netherlands.

Teucrium polium are the most popular. The latter is called in Arabic: "The children's *Teucrium*", due to its popular use for children. The medicine is prepared as tea.

Another common use is for the treatment of colds and coughs. The plant material is prepared as tea and for inhalation as a steam. In these cases a mixture of more than one species is used. Quite unexpected is the fairly wide use in the treatment of heart disorders, dropsy, swellings, rheumatism and localized paralysis. These aspects are mentioned in the literature in only two cases (Table 1).

Common is the external use of plant material in the treatment of wounds and inflammations. The affected area is treated with an aqueous extract of leaves, with liquid squeezed from leaves, or by just applying on it fresh leaves, mainly of *Salvia fruticosa*, which are wider and softer than the leaves of most of the other species.

Other uses are also mentioned but in connection with only one or a few of the species. For instance: *Thymus capitatus* and *Satureja thymbra* for the treatment of swellings, mainly in the legs. A steam bath and the application of leaves on the swollen area are recommended. Toothache is treated by chewing leaves of *Satureja thymbra* and *Majorana syriaca*. *Micromeria fruticosa* is considered an important remedy for high blood pressure.

Some of the findings of the present survey are unique to the local area. For instance, *Micromeria myrtifolia*, which is not mentioned in the literature, and *Satureja thymbra*, which is rarely mentioned, are both widely used in the local folk medicine. On the other hand, some of the medicinal aspects listed in the literature, especially in the case of *Majorana syriaca* and *Teucrium polium* (11), were not reported by the informants in our survey.

Four species of *Mentha*, known as medicinal plants in the literature (11) grow wild in Israel; however, they are hardly used locally in folk medicine. The same is true for *Melissa officinalis* and *Marrubium vulgare*. This finding is quite surprising, since these species have been very well known in folk medicine since early times (6).

A survey of the relevant chemical literature showed a high content of essential oils, rich in phenolic compounds, the most important being thymol and carvacrol. Relevant reports on *Majorana syriaca* are by Zaitschek and Levontin (28), Granger et al. (13) and Fliescher et al. (9); on *Micromeria fruticosa* by Bellino et al. (2), on other *Micromeria* species by Phokas et

Table 1. Uses of seven aromatic plants, viz. *Micromeria fruticosa*, *Micromeria myrtifolia*, *Thymus capitatus*, *Salvia fruticosa*, *Teucrium polium*, *Satureja thymbra*, *Majorana syriaca*, widely utilized in the folk medicine of Israel. Information is acquired through interviewing and literature survey.

	Micromeria fruticosa	*Micromeria myrtifolia*	*Thymus capitatus*	*Salvia fruticosa*	*Teucrium polium*	*Satureja thymbra*	*Majorana syriaca*
Indigestion	2(20)	2*	2(1,21)	4(14)	4(4,8,14,16)	2(28)	2(10,28)
Coughs & colds	2(20)	2	(21)	4(28)	(1,4,16)	2	2(16,28)
High blood pressure	2						
Heart disorders	(15)	2	2	3(21)			2(21)
Swellings & dropsy			2	(21)	(5)	2	(21)
External wounds	2		(11,25)	3	2(5)	2	(28)
Headaches		1		2	(10)	1	(21)
Toothache			(21)			2	2
Women's complaints					2(5)		(21)
Worms			(21)		(21)		2
Nervous anxiety		1				1	(21)
Rheumatism		1	(21)	2		1	
Local paralysis			2			1	
External inflammations		1	(21)		1(3,27)		(28)

*Number of independent informants: 4 = Nine or more
3 = Six to eight
2 = Three to five
1 = Fewer than three
( ) = Corresponding literature references

al. (19); on *Thymus capitatus* by Zaitschek and Levontin (28), Rosengarten (22) and Sendra and Cuñat (23); and on *Teucrium polium* by Skrubis (24) and Wassel and Ahmed (27).

It has been shown in many cases that essential oils possess antimicrobial activity (26). Essential oil extracted from *Satureja* species showed antibacterial activity (17, 18). These activities provide a partial explanation for attributing medicinal properties to these aromatic species.

Many of the survey findings are similar to those reported for other species growing in Greece, Italy (11), India, Pakistan (7) and the Middle East (8). However, each of the local healers is familiar with only a limited number of uses and with only one or two of the listed species. Most of the uses are common to all of the ethnic groups - Arabs, Bedouins and Druse - with the exception of *Micromeria myrtifolia*, which is used exclusively by the last group.

Based on our survey it is concluded that all these species should be considered as potentially valuable crops. Further research is needed to develop the appropriate agrotechnical procedures. The most promising species are *Salvia fruticosa, Micromeria fruticosa* and *Majorana syriaca*.

REFERENCES

1. Bailey Y, Denin A. 1975. Desert Plants in the Bedouins' Life. Notes on the Subject of Bedouins (in Hebrew) 5. 2-48 Sede Boquer Publ., Israel.
2. Bellino A, Venturella P, Marceno C. 1980. *Fitoterapia* 51:163.
3. Chopra RW, Nayar SL, Chopra IC. 1956. Glossary of Indian Medicinal Plants. New Delhi, Council of Scientific and Industrial Research.
4. Crowfoot GM, Baldensperger L. 1932. From Cedar to Hyssop. London, The Sheldon Press.
5. Culpeper N. 1943. Complete Herbal. London, Fousham and Co. Publ.
6. Dafni A. 1980. Plant Folklore in Israel (in Hebrew). Haifa, Gastlit.
7. Dastur JF. 1970. Medicinal Plants of India and Pakistan. Bombay, D. C. Taraporevala Sons & Co.
8. Fahmy IR. 1956. *Lebanese Pharmaceut. J.* 4:12.
9. Fliescher A, Snir N, Fliescher G, Yofe A. 1980. The secret of the biblical hyssop (in Hebrew). *Mada* 24:272.
10. Friedman A. 1966. Folk medicine of the Eastern Jews in the Galile (in Hebrew). M.Sc. Thesis, School of Pharmacy, The Hebrew University of Jerusalem, Israel.
11. Grieve M. 1974. A Modern Herbal. New York, Dover Publ.
12. Guenther E. 1949. The Essential Oils. Vols. I-VI. New York, D. van Nostrand Co. Inc.
13. Granger R, Passet J, Lamy J. 1975. *Rivista Italiana EPPOS* 57:446.
14. Hareuveni A. 1930. Medicinal and Magic Plants of the Arabs in Israel (in Hebrew). *Harefua* 4:113.
15. Kohen A. 1935. Wild plants used by People (in Hebrew). *Teva vaAretz* 3: 217.

16. Levy S. 1978. Medicine Hygiene and Health of the Bedouins of South Sinai. Publ. Soc. of Nature Preservation, Tel Aviv, Israel.
17. Pelleceur J, Jacob M, Simeon de Buochberg M, Allegrini J. 1979. *Planta Med.* 36:256.
18. Pelleceur J, Jacob M, Simeon de Buochberg M, Dusart G, Attisso M, Barthez M, Gourgas L, Pascal B, Tomei R. 1980. *Pl. Med. Phytothér.* 14:83.
19. Phokas G, Patouha-Volioti G, Katsiotis S. 1980. *Pl. Med. Phytothér*: 14: 159.
20. Reshef Y. 1965. Comments on an article by Shukari Araf (in Hebrew). *Teva vaAretz* 7:177.
21. Riani Y. 1963. Medicinal Drugs of the Yemenite Jews (in Hebrew). M.Sc. Thesis, School of Pharmacy, The Hebrew University of Jerusalem, Israel.
22. Rosengarten F. 1973. The book of spices. New York, Pyramid Communications, Inc.
23. Sendra JM, Cuñat P. 1980. *Phytochemistry* 19:1513.
24. Skrubis BG. 1972. *Flavour Ind.* 3:566.
25. Tackholn V, Drar M. 1954. Flora of Egypt. Vol. III. Cairo, University Press.
26. Tyler VE, Brady LR, Robbers JE. 1976. Pharmacognosy. Philadelphia, Lea & Febiger.
27. Wassel GM, Ahmed SS. 1974. *Pharmazie* 29:351.
28. Zaitschek D, Levontin S. 1972. Further information about essential oils of Labiatae in Israel containing phenols (in Hebrew). *haRokeah haivri* 14:284.
29. Zohary M. 1976. A New Analytical Flora (in Hebrew). Tel Aviv, Am Oved Publ. Ltd.

NEW PHARMACOLOGICALLY IMPORTANT FLAVONOIDS OF *THYMUS VULGARIS*

C.O. VAN DEN BROUCKE

1. INTRODUCTION

Thyme, *T. vulgaris* L. as well as *T. serpyllum* L. (Labiatae), is commonly used as a cough-medicine. Especially liquid thyme extracts are in some countries a fundamental part of many galenic preparations with antitussive action. In this work, we studied the physiological effectiveness of these preparations and looked for the active substance(s).

Up to now the volatile oil with its phenols has been proposed as the active fraction of thyme. However, the discovery of many chemotypes of *T. vulgaris* with a very low content of phenols (0.0-0.1%) in the oil claim standardization of the plant.

Different authors (4, 5, 6, 8, 11) reported the secretomotoric, secretolytic and disinfectant properties of thyme oil, thymol and carvacrol. De Keuning (2), Lendle and Lü-Fu-Hua (7) proposed that the active compounds of thyme are non-volatile unknown substances. We proved the spasmolytic action of thyme extracts on the smooth muscles of the isolated guinea-pig ileum as well as on those of the guinea-pig trachea (9). Since phenols could not be responsible for the tracheal relaxation and inhibition of the ileum contractions, we looked for other non-volatile active components.

Up to now the only identified flavonoids in *T. vulgaris* L. are apigenine, luteoline, luteoline-7-glucoside and 7-diglucoside luteoline (1).

2. MATERIAL AND METHODS

2.1. Materials

Thymus vulgaris L. used in this investigation was provided by Denolin n.v., Braine-l'Alleud, Belgium.

2.2. Pharmacological methods

Guinea-pigs weighing 300 to 500 g were used. The animals were killed with a blow on the head.

Margaris N, Koedam A and Vokou D (eds.): Aromatic Plants: Basic and Applied Aspects.

ISBN 90-247-2720-0.
Printed in the Netherlands.

2.2.1. Guinea-pig ileum. From the part adjacent to the ileocaecal junction of the ileum a terminal portion of 2.5 cm was isolated. The cleaned preparation was placed vertically in a 20 ml thermostatically controlled organ bath at 37°C, containing Tyrode's solution, bubbled with 95% O_2 and 5% CO_2. Contraction of the preparation was recorded by an isotonic smooth muscle transducer (Harvard Electronic Transducer 386) with a tension of 1 g. Cumulative concentration-response curves to carbachol, histamine, DMPP (1,1 dimethyl-4-phenyl-piperaziniumiodid) and $BaCl_2$ were obtained according to Van Rossum (10). After two standard dose-response curves with agonists were obtained, the guinea-pig ileum was incubated with the antagonist drugs for 5 min. The dose-response curve of the agonists were then redetermined while the antagonist remained in the bath. On each piece of ileum only one concentration of antagonist was tested.

The affinity of a non-competitive antagonist for its receptor pD_2' is the negative logarithm of the molar concentration of the antagonist that produces 50% reduction of the maximum response obtained with the agonist. pD_2' values were calculated by the following equation:

$$pD_2' = pD_x' + \log \frac{E_{Am}}{E_{ABm}} - 1$$

pD_x' is the negative logarithm of the molar concentration of the antagonist in the presence of which the maximum response of the intestinal strip to the agonist is E_{ABm}; E_{Am} is the maximum control response to the agonist.

2.2.2. Guinea-pig trachea. The zig-zag tracheal strip was used according to the method of Emmerson and Mackay (3). The preparation was set up in a 20 ml organ bath at 37°C, containing Krebs solution 95% O_2 and 5% CO_2. Responses were recorded isometrically by a Harvard muscle transducer 363. Tracheal tone was obtained by carbachol 0.5 x 10^{-6}M. Drugs were added cumulatively to the bath until no further relaxation was seen. At the end of recording of each dose-response curve, a dose of DL-isoprenaline producing a maximum relaxation of the preparation was added. Responses to applied drugs were expressed as a percentage of this maximum relaxation. pD_2 is the negative logarithm of the molar concentration of agonist that produces 50% relaxation of the trachea. The ratio of the maximum relaxation by the tested drug to the maximum response to DL-isoprenaline was expressed as intrinsic activity. Relative potency was determined with respect to papaverine.

2.3. Drugs used

Carbachol chloride, histamine dihydrochloride, DMPP (1,1-dimethyl-4-phenyl-piperaziniumiodide), $BaCl_2$, DL-isoprenaline, apigenine, luteoline, thymol carvacrol and papaverine.

3. RESULTS

3.1. Chemical isolation of new flavonoids

Dried and milled leaves and flowering tops of *T. vulgaris* were extracted continuously with petroleum-ether (bp 40-60^{0}C). The defatted plant material was extracted with methanol and concentrated. After precipitation with ether, the methanol-ether fraction was partly evaporated and digerated with hot water. After removal of the precipitate and evaporation of the methanol, the aqueous layer was partitioned with ether. This ether fraction was evaporated and gel chromatography over Sephadex LH 20 yielded active fractions. Column chromatography over silicagel with chloroform as eluent resulted in three components 30, 3B and 2B, purified by preparative thin layer chromatography (Fig. 1). The complete structure of these three flavonoids will be elucidated in the near future.

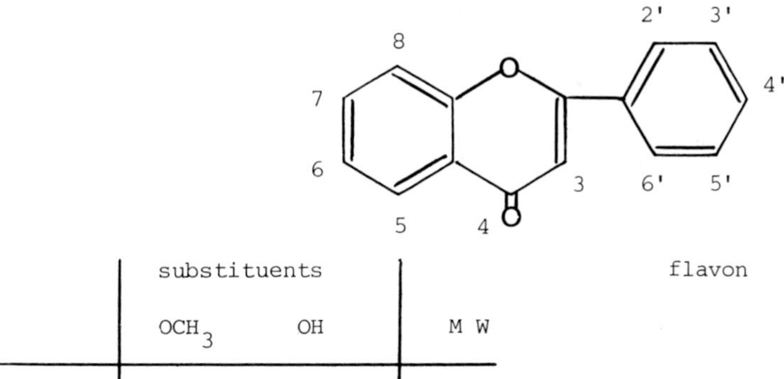

	substituents		
	OCH_3	OH	M W
2 B	3	3	360
3 O	3	2	344
3 B	4	2	374

FIGURE 1. Isolated flavonoids from *T. vulgaris*.

3.2. Pharmacological properties of 30, 3B and 2B.

3.2.1. Guinea-pig ileum. Cumulative dose-response curves for carbachol, histamine, DMPP and $BaCl_2$ were established in the presence of thymol, carvacrol, apigenine, luteoline, 30, 3B, 2B and papaverine. Fig. 2 shows the average dose-response curves obtained for carbachol with 30 as antagonist. All tested drugs caused a dose dependent depression of the maximum responses to carbachol, histamine, DMPP and $BaCl_2$ without any shift along the dose axis. These results suggest that they antagonize these agonists by a non-competitive mechanism of action. The pD_2' values calculated from the maximum depressions of the curves are shown in Table 1. The spasmolytic action of thymol and carvacrol is significantly less potent than that of the flavonoids.

3.2.2. Guinea-pig trachea. All tested drugs exhibited a dose-dependent relaxation of the carbachol induced tracheal strip. Fig. 3 illustrates the effect of cumulative addition of thymol, carvacrol, apigenine, luteoline, 30, 3B, 2B, papaverine and DL-isoprenaline.

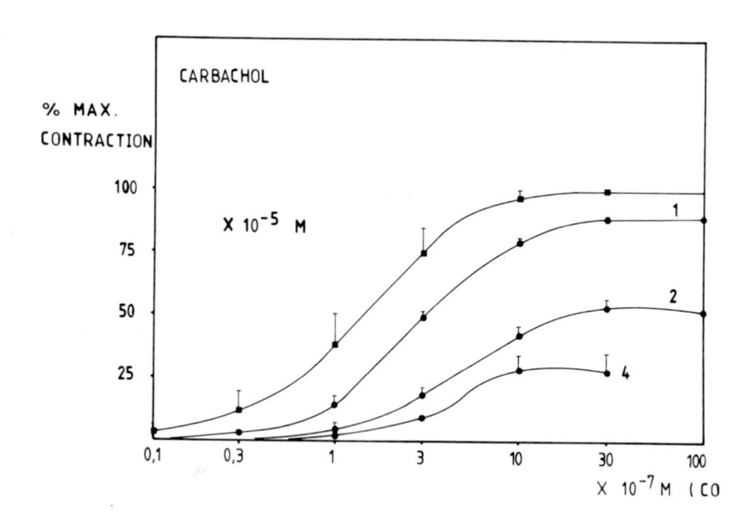

FIGURE 2. Cumulative dose-response curves for carbachol on the guinea-pig ileum. (●) carbachol in the presence of compound 30 in different concentration (1,2 and 4 X 10^{-5} M). Each point represents the mean of 4 values ± SE"; (■) carbachol control curve (mean of 94 curves).

TABLE 1. pD_2' values on the guinea-pig ileum. In parentheses the number of individual curves for the calculation of pD_2' values.

Antagonist \ Agonist	pD_2' on guinea-pig ileum			
	Carbachol	Histamine	D M P P	$BaCl_2$
Thymol	3.70 ± 0.14 (7)	3.72 ± 0.02 (3)	4.74 ± 0.14 (7)	4.13 ± 0.14 (3)
Carvacrol	3.73 ± 0.25 (4)	3.64 ± 0.08 (4)	4.86 ± 0.20 (6)	4.14 ± 0.08 (3)
Luteoline	4.76 ± 0.10 (5)	4.74 ± 0.14 (3)	4.95 ± 0.06 (4)	4.52 ± 0.14 (8)
Apigenine	4.75 ± 0.23 (8)	5.00 ± 0.10 (3)	4.96 ± 0.15 (4)	4.46 ± 0.17 (7)
3 B	4.46 ± 0.23 (5)	4.60 ± 0.12 (8)	5.00 ± 0.26 (5)	4.14 ± 0.19 (4)
2 B	4.55 ± 0.12 (4)	4.73 ± 0.04 (4)	4.66 ± 0.04 (4)	5.02 ± 0.22 (8)
3 O	4.69 ± 0.10 (7)	4.54 ± 0.15 (7)	5.13 ± 0.18 (7)	4.25 ± 0.07 (5)
Papaverine	5.03 ± 0.16 (8)	4.98 ± 0.12 (11)	5.18 ± 0.24 (8)	4.62 ± 0.24 (8)

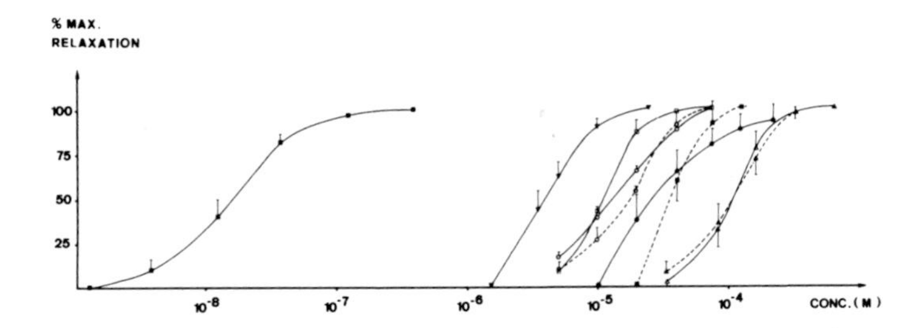

FIGURE 3. Cumulative log dose-response curves for DL-isoprenaline (—■), papaverine (—▼), 3O (—□), 2B (----o), 3B (—o), luteoline (—●), apigenine (----■), thymol (—△) and carvacrol (----▲) on the isolated guinea-pig trachea. Each point represents the mean +SD of four values.

TABLE 2. pD_2 values, intrinsic activity and relative potency to papaverine.

Agonist	pD_2 on guinea-pig trachea	Intrinsic activity	Relative potency (papaverine = 1)
Carvacrol	3.95 ± 0.11	1.0	0.036
Thymol	3.96 ± 0.08	1.0	0.037
Apigenine	4.43 ± 0.04	1.0	0.108
Luteoline	4.55 ± 0.12	0.92	0.143
2 B	4.74 ± 0.10	1.0	0.222
3 B	4.88 ± 0.01	1.0	0.30
3 O	4.95 ± 0.02	1.0	0.357
Papaverine	5.40 ± 0.08	1.0	1.0
DL-Isoprenaline	7.80 ± 0.11	1.0	253

REFERENCES

1. Awe W, Schaller JF, Kümmel HJ. 1959. *Naturwissenschaften* 46:558.
2. De Keuning JC. 1951. Thesis, Utrecht.
3. Emmerson J, Mackay D. 1979. *J. Pharm. Pharmacol.* 31:798.
4. Freytag A. 1933. *Pflügers Arch.* 232:346.
5. Gordonoff T, Janett F. 1931. *Z. Exper. Med.* 79:486.
6. Gordonoff T, Merz H. 1931. *Klin. Wschr.* 10:928.
7. Lendle L, Lü-Fu-Hua, 1937. *Schmied. Arch.* 184:89.
8. Schilf E. 1932. *Schmied. Arch.* 166:22.
9. Van Den Broucke CO, Lemli JA. 1981. *Planta Med.* 41:129.
10. Van Rossum JM. 1963. *Arch. Int. Pharmacodyn.* 143:299.
11. Vollmer H. 1932. *Klin. Wschr.* 11:590.

PLANT SYSTEMATIC INDEX

CHEMICAL INDEX